Special Publication No. 47

Recent Advances in the Chemistry of Meat

The Proceedings of a Symposium Organised by the Food Chemistry Group of The Royal Society of Chemistry and the Food Group of The Society of Chemical Industry

ARC Meat Research Institute, Langford, Bristol
14th—15th April 1983

Edited by
Allen J. Bailey
ARC Meat Research Institute

The Royal Society of Chemistry
Burlington House, London W1V 0BN

British Library Cataloguing in Publication Data

Recent advances in the chemistry of meat —
  (Special publication/Royal Society of Chemistry,
  ISSN 0260-6291;47)
  1. Meat
  I. Bailey, A. J.   II. Series
  641.3'6   TX373

  ISBN 0-85186-905-X

ML 1563

Printed in Great Britain by
Whitstable Litho Ltd., Whitstable, Kent

Introduction

Structural and biochemical differences in the muscle fibre and connective tissue components of muscle types evolved for different physiological functions lead to considerable differences in the eating quality of the muscles as meat. In particular, muscle differentiation causes variation in meat texture and flavour. An understanding of the components of muscle, and the conversion of muscle to meat, is a prerequisite for predicting and controlling this variability.

A background knowledge of meat structure is also important in identifying and subsequently correcting technical problems in the meat industry. For example, understanding the biochemical and biophysical changes associated with rigor development, and their control, enabled the development of techniques to overcome the problem of cold-shortening toughness. Solving the problem of soft fat in pigmeat, which has arisen more recently, will require fundamental studies across scientific disciplines.

It should be remembered that animal products provide three-quarters of the protein, one-third of the energy and most of the calcium and phosphorus in our diet. Further, the meat industry accounts for one-quarter of the household expenditure on food. Despite the size and importance of the industry, it is still primarily run on traditional lines. However, I believe that our rapidly increasing knowledge of the structure and rigor processes of muscle has now reached the stage where one can ask specific questions of the processes currently controlled on an empirical basis.

It is with this in mind that, on behalf of the Food Chemistry Group of the Royal Society of Chemistry, a meeting was organised in April 1983 at the ARC Meat Research Institute, Langford, where a number of basic scientists were invited to bring us up to date on the chemistry of meat.

Allen J. Bailey

Acknowledgements

I should like to express my special thanks to Miss Davina Bicknell who handled all the arrangements before and during the symposium, and then took on the daunting task of organising the typing of the papers. Without her painstaking attention to detail in the proof-reading of these papers and the expert typing of Miss Joanna Rowlands, the book would probably have never appeared.

Contents

1
The Structure of Muscle and Its Properties as Meat

By C. Lester Davey
MEAT INDUSTRY RESEARCH INSTITUTE OF NEW ZEALAND (INC.), P.O. BOX 617, HAMILTON, NEW ZEALAND

INTRODUCTION

The early studies of mammalian skeletal muscle largely laid the foundations of modern biochemistry and physiology, since the burgeoning scientific curiosity in the middle of last century focused on muscle for diverse examinations. Unlike other biological tissue, skeletal muscle can live on for some time after the animal's death. It can be stimulated to contract and do work, and it has remarkably well ordered and complex repeating structures that are highly amenable to histological and X-ray diffraction analyses. It has the added merit of being a valuable and nutritious food. The contractile machine embodies many of the principles of an internal combustion engine, as the years of successful research into skeletal muscle function have shown. We now know how its activity can be triggered to start and stop and how mechanical action is translated to bodily movement. We know, too, much about its energy resources and their conversion to produce mechanical events; we also know how a muscle discharges its waste products.

This review presents knowledge of the muscle cell, giving particular emphasis to features that determine the toughness of meat. Better knowledge of animal breeding and nutrition, of disease eradication and of microbial control has added to our confidence in meat as a wholesome food. Assuming it can be made so, then toughness is its least-wanted feature. To understand the property of toughness it is necessary to study the components of meat that determine its mechanical strength. The mechanical events associated with the contraction of muscle and the biting of meat have much in common, the forces generated by both being transmitted through the same structural elements. Several reviews have discussed the relationship between muscle and meat; [1-4] the present one brings together more recent information, extending knowledge of the physiology of muscle tissue into the realm of how this tissue serves as food.

THE MUSCLE CELL

The fibrous nature of muscle is evident to all. Muscle cells (fibres) vary greatly in size, from tiny fibres in the muscles of diminutive animals to greatly elongated fibres, 20-30 cm long, in muscles of large animals. The fibrous, or filamentous, character of the muscle cell extends to most of its components as well. Little wonder this is so, since it is hard to imagine muscular contraction happening in any other way than through the integrated development of tension in fibrillar elements with distinct tensile properties.

Each muscle cell is surrounded by a collagenous reticular layer of endomysium, beneath which is the cell membrane or sarcolemma. The sarcolemma is triple-layered and about 10 nm thick.[5] It transmits nervous signals along the surface of the cell from motor end-plates and also maintains the cell in delicate osmotic balance. The sarcolemma probably makes little contribution to the tensile properties of the muscle fibre, from which it can be peeled leaving the structural components of the cell intact.[6]

The skeletal muscles of an animal differ in colour, ranging from red through pink to white.[7,8] So-called red muscles tend to be slow-acting and show considerable stamina. In contrast, white muscles tend to be relatively fast-acting and are easily fatigued. Colour is largely related to the concentration of myoglobin in the muscle. Independent of muscle function, the concentration of myoglobin is usually low in rapidly respiring small animals such as the rabbit (0.2% wet weight of muscle) and increases to 0.7% or more of the muscle weight in horses.[9] Much higher levels (5-8%) are found in the muscles of seals and some species of deep-living whale, presumably to support sustained underwater pursuits.[10]

The individual muscle cells are also loosely classified as red, intermediate and white, and most muscles contain a mixture of these types. Many structural and functional distinctions can be made amongst muscle-cell types. Consider the extremes: in comparison with white muscle fibres, the red ones have a smaller diameter, have structurally looser and simpler neuromuscular junctions, are richer in mitochondria and therefore are more reliant on oxidative metabolism. The distinction extends to fine aspects of ultrastructure and to the isoenzymes; for instance, red muscle fibres have a higher proportion of slow myosin, which exhibits a relatively low ATPase activity.[11]

Within the muscle cell are a number of distinct and significant structures: the contractile apparatus, the elucidation of which owes much to the muscle biologists; a refractory cytoskeletal framework, the definition of which is still in progress and owes much to the meat scientist; and a reticulum of sarcoplasmic tubules and vesicles, the understanding of which owes something to scientists in both disciplines.

Some 80% of the muscle-cell volume is taken up by the contractile apparatus or myofibrils – threadlike elements, each approximately 1 μm in diameter, that run the length of the muscle cell. Surrounding the myofibrils is the enzymic pool, the sarcoplasm, and lying in this are the various discrete cell components – the nuclei, responsible for cell morphology and differentiation; the mitochondria, responsible for oxidative metabolism; the cytoskeletal framework, responsible for holding the contractile apparatus together; and the sarcoplasmic reticulum, responsible for transmitting nervous signals and generating Ca^{2+} fluxes for excitation-contraction coupling and for relaxation. Then there are the glycogen granules, serving as the source of much of the cell's metabolic and mechanical energy, without which the cell would die.[12]

The contractile apparatus

The myofibrils contain filaments, which are assembled to form a highly ordered, repeating structure familiar to the muscle investigator since the early days of microscopy. Cross-striations are seen in longitudinal sections of muscle even at quite low magnification, since the repeating units are normally in register within and between the myofibrils in the cell. In uncontracted muscles the striations appear under phase-contrast microscopy as a spaced, repeating sequence of light, isotropic I-bands and dark, anisotropic A-bands. Transverse Z-discs bisect the I-bands, and relatively light H-zones bisect the A-bands. Dark M-lines can also sometimes be observed bisecting the H-zones and, even less distinctly, N-lines bisecting the I-bands on each side of the Z-discs. The present stage of our knowledge of the fine structure of muscle has been the subject of recent reviews.[13],[14]

The contractile unit within the repeating assembly is the myofibrillar sarcomere, bounded at each end by a Z-disc (Fig.1). Recent research[15] has identified up to 20 different proteins within the sarcomere, although the two major ones, myosin and actin, make up 65% of the total quantity and together are recognised as the basic components of the contractile apparatus. Morphologically, the contractile apparatus is organised into two sets of filaments running parallel to the muscle fibre. Thick myosin filaments, approximately 12 nm in diameter and having a length of 1500 nm, occupy the central part of the sarcomere. Thin F-actin filaments, 8 nm in diameter and 1000 nm in length, are attached to the Z-discs, much as the bristles of a brush are attached to its handle or spine.

There is now almost universal agreement on the correctness of what has become known as the sliding-filament theory of muscular contraction.[16] Contraction is regarded as being due to a relative sliding of the thick and thin filaments in the sarcomere. With contraction of the sarcomere, the isotropic I-bands shorten as the thin filaments are drawn into the spaces between the thick filaments of the anisotropic A-bands. Thus, what appears as contraction at the level of the whole muscle or at the level of the sarcomere does not have a counterpart in terms of a shrinkage in, or contraction of, the molecular components of the contractile apparatus. It is the interaction of the thick myosin-containing filaments with the thin, actin-containing filaments in their regions of overlap that causes shortening and generates the force of muscular contraction.

The thick filaments have an important and distinctive feature, in that except for a smooth central zone (160 nm) they contain regularly spaced protruberances or cross-bridges along their length.[16] The protruberances are, in fact, the head ends of individual myosin molecules, the tail sections of which are aggregated in parallel into the underlying thick-filament shaft. And so the filamentous structure of muscle exists through to the molecular level. Individual myosin molecules can be extracted at high salt concentration and visualised by electron microscopy. They are double headed (each head is about 9 nm long and 4 nm wide) and have a highly helical, doubly wound tail.

The precise way in which the myosin molecules are stacked is a matter of speculation. However, from direct electron-microscopic observation and more especially from analyses by low-angle X-ray diffraction, considerable structural detail is revealed.[16] In the 160 nm central zone of the thick filament, myosin molecules are stacked in an ordered, anti-parallel fashion to give a smooth area consisting solely of myosin tails. In

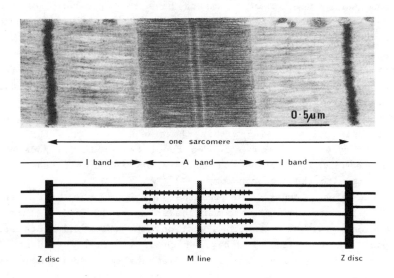

Figure 1 The muscle sarcomere stretched to best display fine details of its ultrastructure. The I-bands consisting of thin filaments which occur either side of the Z-disc, beyond the sides of the micrograph, and similarly in the diagrammatic representation, to the A-bands of the adjacent sarcomeres. The regularly spaced protruberances on the thick filaments are the heads of individual myosin molecules. They are not visible in the micrograph. The N-lines that run across the I-bands on each side of the Z-disc have also not been resolved in the micrograph.

contrast, the molecules on each side of the smooth zone are stacked in essentially parallel fashion, the myosin heads protruding from the thick filaments at intervals of 14.3 nm. The protruberances at each level are equally spaced around the filament, either as three or as four sets of double heads. The protruberances are not in line along the thick filaments; rather, through progressive axial displacement from one 14.3 nm level to the next, they trace a helical pathway. If there are three protruberances at each level, their helical disposition is three-stranded, and if four, then their disposition is four-stranded. To what extent head disposition is important to the contraction mechanism is as yet unknown.[16]

The main protein of the thin filaments is actin, which, after myosin, is the second most common of the 20 or so proteins of the sarcomere. Thin filaments have the appearance of two long-pitch helical strands of linked actin subunits wound round each other.[17] By the nature of its construction, this double strand of globular actin molecules probably does not have much tensile strength. Thin filaments probably gain strength from the other components they contain. Located at paired intervals (38.5 nm) along the two helical grooves of the double strands of actin are globular troponin complexes (Troponins T, C and I) linked by tropomyosin strands lying within these grooves. Thus the thin filament resembles a four-stranded rope: two of the strands are of linked globular actins and two are of tropomyosin beaded at intervals with the troponin complex.[18]

The highly ordered longitudinal arrangement of thick and thin filaments in a sarcomere also extends to a three-dimensional lattice. In cross-section, the thick filaments are disposed in an hexagonal array, which is especially evident at the level of the M-line, where the filaments are held together by interconnections of myomesin.[19] An ordered arrangement of the thick and thin filaments is also clearly demonstrated in the region of filament overlap, with six thin filaments surrounding each thick filament. In contrast, the thin filaments in the region of the I-band are less regularly dispersed.

The cytoskeletal framework

The purpose of this review is to relate the structure and function of muscle tissue to causes of toughness in meat. In this regard there is another important set of proteins within the muscle cell. Evidence has recently accumulated that in addition to the thick- and thin-filament array, a fibrous cytoskeletal structure exists around and within the myofibrils of the muscle cell.[20,21]

It has long been recognised that the simple view that sarcomeres contain mainly two sets of overlapping filaments presents a number of doubts and problems. On selective extraction of myosin, 'ghost' myofibrils remain resiliently intact.[22] Other experiments have shown that thick filaments stay anchored steadfastly at the centre of the sarcomere during active contraction or passive stretching. A third set of fine filaments especially prominent in insect flight muscle has been observed running between the Z-discs,[23] and in the last five years or so increasing attention has been given to them, as well as to other distinctive cytoskeletal material that appears to hold the sarcomere together in its regular order.[20,24,25] For example, the cytoskeleton has a distinctly lateral component linking myofibrils at the Z-discs so that adjacent myofibrils stretch and contract in reasonable concert. We therefore have a view of a cytoskeleton that remains after the removal of myosin and actin - a residue with major components running between the Z-discs either along or perpendicular to the muscle cell direction. Such a network is bound to be a complex of interacting proteins.[15]

Of all the cytoskeletal proteins, connectin (or titin) is the most prevalent, accounting for 6% of the myofibrillar bulk. Connectin apparently has a most important role in preserving the complex integrity of the sarcomere through all the stresses and strains imposed on it with the mechanical working of the muscle cell. Because of its insolubility and apparent lack of chemical markers, this protein (or proteins) is difficult to characterise (mol.wt. 700,000-1,000,000 dalton). Antibody studies show that connectin is located throughout the sarcomere, with the possible exception of the Z-disc. It also appears to lie longitudinally between the myofibrils. Demonstration of the detailed structure of this protein will require that it be highly purified and well characterised - a difficult task given its refractory nature.

Connectin is the most likely candidate for the largely disregarded fine filaments seen to straddle the sarcomere once the thick and thin filaments have been removed. These fine filaments (diameter $\simeq$ 2 nm), which were recently re-discovered after myosin and actin extraction and in muscles stretched beyond the thick- and thin-filament overlap,[26] have been termed g-filaments. In all their major properties they are identical, or very closely similar, to connectin. According to Locker,[27] g-filaments are centred on the Z-disc, from which they extend on either side to insert into the thick filaments of adjacent sarcomeres. They thus form a cytoskeletal syncytium within the sarcomere. A later section of this chapter discusses the apparent key role of g-filaments in determining the toughness of meat.

Other proteins in lesser quantity have recently been shown to occur in the sarcomere. These, too, are likely to be part of the cytoskeleton. Myomesin (subunit weight 165,000 dalton) is a major component of the M-line and helps hold the thick filaments of the A-band in lateral register.[19,28] Since myomesin binds strongly to myosin, it may well also serve as the condensing substructure for the central anti-parallel packing of myosin molecules in the thick filaments; it may also serve to limit the length of these filaments to 1500 nm.

Many cytoskeletal proteins are present in, or associated with, the Z-disc: α-actinin (200,000 dalton), desmin (55,000 dalton), an as yet unnamed but different protein (55,000 dalton), vimentin (58,000 dalton) and synemin (23,000 dalton). Immuno-fluorescence studies suggest that of these proteins, desmin is located as a net, not within but around the Z-discs, linking adjacent myofibrils. Another high molecular weight protein is the N_2-line protein ($\simeq$60,000 dalton). As the name implies, it is found in the N_2 lines that lie across the I-bands parallel to the Z-discs.[29]

The sarcoplasmic reticulum

The third structural entity of great importance is the complex system of longitudinal vesicles and transverse tubules which wraps around each myofibril and which collectively is known as the sarcoplasmic reticulum. The transverse tubules originate as invaginations of the sarcolemma and carry the triggering signal for muscular contraction. The longitudinal vesicles, although more apparent, are more difficult to envisage. They totally envelop the myofibrils, and when prompted by the signal from the transverse tubules they discharge Ca^{2+} to the neighbouring myofibrillar domain to induce contraction. This very simple view of what is a complex Ca^{2+} storage and pumping system is described in detail in the excellent review of Ebashi.[30]

Muscular contraction

The relative sliding of thick and thin filaments in the sarcomere is the mechanical basis of muscular contraction.[31,32] The cyclical attachment of thick-filament protruberances (cross-bridges) to the globular actin units of the thin filaments produces the relative sliding of filaments, leading to tension development in and contraction of the whole muscle. In live mammalian muscle at rest, this cyclical interaction is not occurring; the myosin heads are not in combination with actin, and therefore the muscle is freely and reversibly extensible.[16]

It is the release of Ca^{2+} from the sarcoplasmic reticulum on nervous stimulation that triggers cyclical cross-bridging. In the simplest of terms, when the troponin complex in the thin filaments receives Ca^{2+}, the tropomyosin strands tighten, which allows the swivelling heads of myosin access and attachment to actin.[16] Then, through dephosphorylation of ATP bonded on the heads of myosin molecules, free energy is released for relative sliding between thick and thin filaments. The making and breaking of such bridges continues for as long as a pool of ATP remains, or until the sarcoplasmic reticulum retrieves its released Ca^{2+}.[30]

With this concept of contraction, we are close to understanding how the free energy released from the dephosphorylation of one molecule of ATP is used to create a quantum of movement at the level of a single cross-bridge. The concept owes much to our knowledge of both the molecular architecture of muscle and the ATP hydrolysing capacity of the contractile apparatus. It allows us to consider cross-bridge action in terms of the sequence of kinetic steps associated with ATP hydrolysis. A highly simplified sequence of possible reaction steps in the cyclical interaction of actin and myosin is shown in Fig.2. For a more comprehensive analysis of these molecular events, reference can be made to Squire.[16]

RIGOR MORTIS

When an animal is slaughtered, its musculature lives on and, although inevitably in the process of dying, remains alive until <u>rigor mortis</u> is finally established. Our knowledge of rigor is based on the early studies of Bate-Smith and Bendall,[33,34], although much information has been added since then. Bendall[12] has reviewed this information in his excellent account of rigor development.

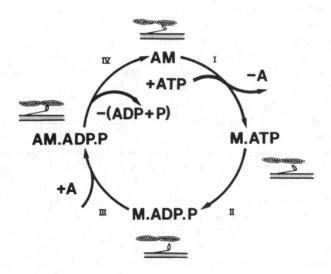

Figure 2 Diagrammatic representation of the cylical force-generating mechanism of muscular contraction. A, actin; M, myosin; AM, cross-bridged actin and myosin. In the cycle ATP binds to myosin, which detaches from actin (Step I). The bound ATP undergoes a presumed readjustment (ADP.P) reflected in the changed angle of the myosin head in relation to the thick-filament axis (Step II). Myosin in this state reattaches to actin (Step III). The ADP.P is dephosphorylated and the myosin head attached to actin moves from a perpendicular to a sharply angled position. This movement induces a quantum of thin-filament movement relative to the thick filament (Step IV).

If well oxygenated and well supplied with the nutrient glycogen, excised muscle strips can remain in an active, pre-rigor condition for many days. They can be stimulated to contract and do work, and are reversibly extensible. If starved of oxygen, they enter rigor mortis some minutes or hours post mortem, becoming rigid and inert.[12]

In living muscle at rest ATP is slowly dephosphorylated to ADP, producing free energy for various metabolic requirements. The slow dephosphorylation continues after death but, in the absence of oxidative metabolism, resynthesis of ATP cannot long keep pace with breakdown. Thus, after a delay, the ATP concentration begins to fall and the muscle enters the rapid phase of rigor development, which continues until virtually all the ATP has disappeared. During the earlier delay phase, the thick and thin filaments are free to slide past each other, so the muscle is extensible and can be stimulated to contract. During the rapid phase, however, myosin cross-bridges become established between adjacent thick and thin filaments in the region of their overlap. With the final disappearance of ATP, cross-bridging is complete and the muscle has entered the inert state of rigor mortis.[12] In so doing it has been converted into meat.

Cold shortening

Of great interest to the meat scientist is the fact that muscles can enter rigor mortis with different degrees of overlap between the thick and thin filaments. In other words, muscles can be set in rigor at different degrees of stretching or shortening. Carcass posture, for instance, will determine to what degree muscles will have been passively stretched or can be shortened on the skeleton.[35,36] Of even more significance, pre-rigor muscles of the main domestic meat species can be induced by chilling temperatures to enter a contracture, in which state they can be set in rigor mortis.[37] As an instance, when unrestrained pre-rigor ox M. sternomandibularis is held at about 2°C for 24 hr, it shortens by up to 50% of its excised length. The magnitude of this type of contracture, called cold shortening, increases with falling temperature and is reduced if chilling is delayed. Compared with physiological contraction, cold shortening occurs extremely slowly and develops only about

5% of the power of a maximal tetanic contraction. However, in most respects the phenomenon is similar to normal contraction, in that in involves ATP dephosphorylation. It occurs in beef, sheep and turkeys; it is also present in pigs, but with less intensity, and is confined to the red muscles of the rabbit.[2]

To what extent can cold shortening occur in a carcass when most of the skeletal attachments are intact? A few muscles such as the neck muscles are severed during carcass dressing and therefore can shorten. Other muscles, notably the M. longissimus, remain attached to the skeleton, but since most of their constituent fibres insert into flexible epimysia, these muscles, too, can shorten. Even muscles that are fixed absolutely at both ends can cold shorten over a part of their length if portions are subjected to different chilling rates. This has been demonstrated in restrained ox M. sternomandibularis insulated at both ends and held at 2°C. A zone of marked shortening occurred in the rapidly chilled central region with concomitant stretching at the two insulated ends.[38]

MUSCLE AGEING - THE RESOLUTION OF RIGOR MORTIS

Although rigor development is the most pronounced post-mortem change that muscle undergoes, other, slower-acting changes can occur, as shown by changes in the muscle's physical characteristics. With time, the rigor muscle reverts from its rigid condition to become more flaccid. The terms 'resolution of rigor' or meat 'ageing' have been coined to describe this post-mortem change. Ageing is the most distinctive of the physical changes that occur after rigor development and can be quantified by measuring the load-extension characteristics of a muscle along the fibre direction.[39] Pre-rigor muscle readily stretches under light stress (5-15 kPa) and the extension is reversible with little hysteresis. Rigor muscle is refractory and only approaches full and irreversible extension at stresses of more than 200 kPa. In contrast, aged muscle, independent of its stage of contraction, is readily stretched under the same light stresses as is the case with pre-rigor muscle. But with aged muscle the extension is not reversible - the muscle remains extended after load removal. The limit to extension (twice equilibrium length) is, however, virtually the same in all three cases, indicating that connective tissue is the load bearer at full stretch.[40]

Several clear morphological changes appear to be associated with ageing. The Z-discs become disorganised, progressively fading beyond the resolving limit of the microscope;[41] the g-filaments weaken and decay;[42] and lateral adhesion between adjacent myofibrils is lost.[43] Such post-mortem changes can be readily envisaged in terms of the structural components of muscle, and it appears that these changes largely relate to the cytoskeletal fabric. The Z-disc decay is probably a consequence of proteolytic destruction of the ordered α-actinin syncytium that anchors the ends of the actin filaments. The g-filaments are also susceptible to attack by endogenous proteolytic enzymes and virtually disappear from muscle during ageing.[44] The loss of lateral adhesion between myofibrils is again a consequence of proteolytic action on the network of desmin that links adjacent myofibrils at the level of the Z-discs.[20,45]

MUSCLE STRUCTURE AND MEAT TOUGHENING

Mechanical shear devices (tenderometers) are used to measure the toughness of meat. Toughness is measured as the force required to cleave a standard cross-sectional area of cooked meat across the muscle fibres, and the resulting shear measurements are often given as shear-force (SF) values.

When a tenderometer cleaves meat across the muscle fibres, the strain is widely dispersed away from the biting wedge. From analysis of the resulting distortion patterns (Fig.3), longitudinal slipping or shearing between muscle fibres is presumed to develop, and thereafter strain is taken up by the connective tissue. Shearing occurs between myofibrils too, and this strain will be absorbed by the transverse linkages between Z-discs of adjacent myofibrils. The more widely the wedge loading is translated amongst muscle fibres through transverse adhesions, the greater the load must be to bring meat to its yield point. In other words, toughness will depend both on muscle-fibre tensile strength and on how widely transverse linkages disperse the wedge load through the meat sample.[4]

Caution is called for in interpreting the toughness of meat in terms of the fine structure of raw muscle, since the ravaging effects of cooking intervene before toughness is normally measured. Meat is rare if cooked to an internal temperature of

60°C and well done if cooked to about 80°C. During cooking two
distinctly separate phases of toughening develop, as measured by
a tenderometer.[46] The first phase, between 40 and 50°C, is
associated with a loss of myosin solubility and is presumed to
indicate protein denaturation in the contractile system. The
second phase, between about 65 and 75°C, is associated with a 25-
30% shrinkage along the muscle fibres and is largely due to
denaturation of connective tissue proteins. If cooking is
prolonged at about 80°C, toughness decreases due to the
progressive melting of collagen.

Added to the effects of cooking on meat tenderness are the
well known observations that whereas ageing tenderises meat, cold
shortening can indeed greatly toughen it, to the extent that at
high shortenings the tenderising of ageing can be completely
overriden.[47] From these rather disparate and very briefly stated
observations a working theory of toughness in relation to muscle

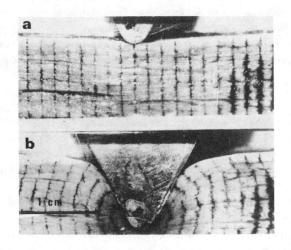

Figure 3 Distortion patterns developed on cleaving cooked ox M.
sternomandibularis; (a) start of wedge penetration; (b) at about
70% wedge penetration.

structure can be developed.

On heating beyond the first phase to 60°C, muscle proteins denature and congeal. The meat, with its reinforcing meshes of cytoskeletal g-filaments and desmin and its connective tissue, disperses a given strain away from the biting wedge, increasing the loading needed to reach yield point. The toughening of the second cooking phase, between 65 and 75°C, can be regarded as an extension of these events. With the shrinkage of the collagen in the connective tissue the tensioning of the reinforcing connective tissue mesh will increase the tensile strength of the meat in much the same way as pre-stressing strengthens concrete. On prolonged cooking at 80°C collagen melts and shear force values revert to about those achieved with 60°C cooking, as would be expected from this concept of toughness.

Locker[27] gives a central role to the g-filaments in his theory of these cooking events. From both tensile-strength measurements and electron-microscopic analyses of meat cooked at different temperatures, he concludes that most of the toughness remaining after prolonged cooking is due to the refractory g-filaments, which determine the tensile strength of the cooked muscle fibre.

As has been stated, the major changes of ageing are the greatly increased ease of muscle extension to twice equilibrium length, the loss of intermyofibrillar adhesion, fading of the Z-discs and a loss of g-filaments. Subtle changes may also occur in the connective tissue,[48] but as the results from prolonged cooking have shown, these do not appear to contribute significantly to tenderisation by ageing. Bringing this information together, it seems likely that ageing occurs through a breakdown of the reinforcing cytoskeletal framework, made up of connectin, including the g-filaments, and desmin. Added support is given to this view by the observations that maximal rates of connectin breakdown and of meat ageing occur at 60°C.[49] The g-filaments are very susceptible to proteolysis; they are readily destroyed by pressure-heat treatment[50] and also during meat ageing.[44] Furthermore, a protein presumed to be desmin is progressively released into a soluble form during meat ageing.[45]

A remaining consideration is what causes the peaked toughness-shortening relationship that is clearly seen in uniformly structured bovine **M. sternomandibularis** (Fig.4). One theory explains the exponential rise in toughness with sarcomere shortening in terms of increased filament overlap, which results in a decreased width of the relatively weak I-bands.[51] However, overall muscle shortening is not a reliable guide to sarcomere length, which varies by as much as 50% for a given shortening.[39] This observation supports that of Voyle[52] who found histologically that fewer than half of the fibres of bovine muscle actively cold shorten, the remainder developing an accommodating crimp. Therefore, the rising phase of toughening may be related not to an equivalent decrease in sarcomere length, but to an increased incidence of highly shortened (1.5 μm or less) sarcomeres, in which the thick filaments are compressed

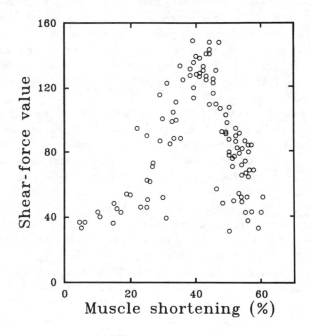

Figure 4 The relationship between the degree of raw muscle shortening and the shear-force value for cooked ox M. sternomandibularis.[38]

against or run through the Z-discs. According to Locker and Wild,[53] as the I-bands progressively disappear with cold shortening, a virtual continuum of g-filaments is established in the shortened muscle which, when cooked, is tough.

The decline in toughness at shortenings above 40% can also be explained readily in terms of raw meat structure. At 40% shortening, those fibres that have contracted are uniformly shortened, with little damage. At higher shortenings, an increasing proportion of sarcomeres are disrupted; fracturing of myofibrils is widespread, and nodes of extreme shortening are interspersed with regions that are stretched to the point of breaking.

In any theory of tenderness it is necessary to consider the importance of not only intermyofibrillar linkages, but also the larger lateral links between muscle cells. On theoretical grounds alone such adhesions would greatly affect meat tenderness. So-called belt-desmosomes link cells, presumably to integrate tension and movement.[54] Adhesive materials such as the fibronectins[55] are present in the sarcolemma, and ageing appears to cause a marked loss in their adhesive properties. As has been mentioned, desmin is the major component of the lateral adhesive links between myofibrils at the Z-discs, and the well observed weakening of these links during ageing will increase the tenderness of cooked meat.

Although much evidence has been assembled demonstrating that toughness relates to the shortened sarcomere, certain contrary observations should be kept in mind. Toughening is not induced in ox M. sternomandibularis that is rigor shortened at 37°C or that is cold shortened and then allowed to enter rigor at 37°C. Lamb carcasses also are not toughened to the extent anticipated by entering rigor at 45°C, despite the fact that rigor shortening would have been considerable.[56] A period of high-temperature holding at some stage in the pre-rigor phase, therefore, appears to overcome toughening, but not shortening.

CONCLUSIONS

The present paper has attempted to link the two sciences of muscle and meat by examining both the structure and the function of muscle as they may affect the toughness of cooked meat. Before an entirely convincing concept of what causes toughness is

developed, a much clearer view of the connectin-based
cytoskeleton is required. As an instance, thick and thin
filaments have dihedral symmetry about the M-lines and Z-discs
respectively, so that the contractile proteins in each half
sarcomere can present themselves to each other in the same
steric sense. Nature undoubtedly imposes the same symmetry
demands upon the various longitudinal and lateral components of
the cytoskeleton. How these demands are satisfied will be the
subject of some daunting unravellings in the years ahead. With
the resolution of the fine structure of the connectin-based
cytoskeleton, molecular biologists must address themselves to the
implications of this structure to the sliding-filament theory.
This theory, itself, is soon likely to be accorded the status of
a scientific reality, so close are our understandings of the
molecular basis of cross-bridge movement.

Until recently it could have been claimed that we were
moving to a satisfactory understanding of how meat toughness
occurs, especially that induced by cold. However, it now appears
that sarcomere shortening per se is not the necessary and
sufficient condition for the development of processing toughness
it was once thought to be.[57,58] New research directions will
need to be identified and theories tested before we can be sure
that tenderness is reliably achieved in meat through effective
process design and control.

RERERENCES

1.   D.E. Goll, M.H. Strommer, D.G. Olsen, W.R. Dayton, A. Suzuki
     and R.M. Robson, "Proc.Meat Ind.Res.Conf.,Chicago", Amer.
     Meat Inst.Found., Arlington, VA, USA, 1974, p.75.
2.   R.H. Locker, C.L. Davey, P.M. Nottingham, D.P. Haughey and
     N.H. Law, Adv.Food Res., 1975, 21, 157.
3.   B.B. Marsh, "Meat, Proc.21st Easter School Agric.Sci.,Univ.
     Nottingham", Butterworths, London, 1974, p.339.
4.   C.L. Davey and R.J. Winger, "Fibrous Proteins: Scientific,
     Industrial and Medical Aspects", Academic Press, London,
     1979, Vol.1, p.97.
5.   R.P. Gould, "The Structure and Function of Muscle", Academic
     Press, London, 2nd Edn, 1973, Vol.II, Part 2, p.186.
6.   R. Natori, Jikeikai Med.J., 1954, 1, 119.
7.   G.F. Gauthier, "The Physiology and Biochemistry of Muscle as

a Food", Univ.Wisc.Press, Madison, USA, 1970, Vol.2, p.103.

8. R.G. Cassens and C.C. Cooper, Adv.Food Res., 1971, 19, 1.

9. R.A. Lawrie, Biochem.J., 1953, 55, 298.

10. J.G. Sharp and B.B. Marsh, "Food Invest.Spec.Report No.58", DSIR, London, 1953.

11. G.F. Gauthier, S. Lowey and A.W. Hobbs, Nature, 1978, 274, 25.

12. J.R. Bendall, "The Structure and Function of Muscle", Academic Press, London, 2nd Edn, 1973, Vol.II, Part 2, p.243.

13. J. Hanson, V. Lednev, E.J. O'Brien and P.M. Bennett, Cold Spring Harbor Symp.Quart.Biol., 1972, 37, 311.

14. J.M. Squire, "Fibrous Proteins: Scientific, Industrial and Medical Aspects", Academic Press, London, 1979, Vol.1, p.27.

15. M.L. Greaser, S. Wang and L.F. Lemanski, Proc.Recip.Meat Conf., 1981, 34, 12.

16. J.M. Squire, "The Structural Basis of Molecular Contraction", Plenum Press, London, 1981.

17. J. Hanson and J. Lowy, J.Mol.Biol., 1963, 6, 46.

18. S. Ebashi, M. Endo and I. Ohtsuki, Q.Rev.Biophys., 1969, 2, 351.

19. T. Masaki and O. Takaiti, J.Biochem., 1974, 75, 367.

20. R.M. Robson, M. Yamaguchi, T.W. Huiatt, F.L. Richardson, J.M. O'Shea, M.K. Hartzer, W.E. Rathbun, P.J. Schreiner, L.E. Kasang, M.H. Stromer, Y.Y.S. Pang, R.R. Evans and J.F. Ridpath, Proc.Recip.Meat Conf., 1981, 34, 5.

21. R.M. Robson and T.W. Huiatt, Proc.36th Recip.Meat Conf., 1983 (In press).

22. J. Hanson and H.E. Huxley, Nature, 1953, 172, 530.

23. M.K. Reedy, "Contract.Muscle Cells Relat.Processes Symp.", Prentice Hall, Englewood Cliffs, 1970, p.229.

24. K. Maruyama, S. Matsubara, R. Natori, Y. Nonomura, S. Kimura, K. Ohashi, F. Murakami, S. Handa and G. Eguchi, J.Biochem., 1977, 82, 317.

25. E. Lazarides, Nature, 1980, 283, 249.

26. R.H. Locker and N.G. Leet, J.Ultrastruct.Res., 1976, 55, 157.

27. R.H. Locker, Proc.Recip.Meat Conf., 1982, 35, 92.

28. J. Trinick and S. Lowey, J.Mol.Biol., 1977, 113, 343.

29. K. Wang and C.L. Williamson, Proc.Nat.Acad.Sci.(USA), 1980, 77, 3254.

30. S. Ebashi, Ann.Rev.Physiol., 1976, 38, 293.

31. A.F. Huxley and R. Niedergerke, Nature, 1954, 173, 971.

32. H.E. Huxley and J. Hanson, Nature, 1954, 173, 973.

33. E.C. Bate-Smith and J.R. Bendall, J.Physiol., 1947, 106, 177.

34. E.C. Bate-Smith and J.R. Bendall, J Physiol., 1949, 110, 47.

35. H.K. Herring, R.G Cassens, G.G. Suess, V.H. Brungardt and E.J. Briskey, J.Food Sci., 1967, 32, 317.

36. C.L. Davey and K.V. Gilbert, J.Food Technol., 1973, 8, 445.

37. R.H. Locker and C.J. Hagyard, J.Sci.Food Agric., 1963, 14, 787.

38. B.B. Marsh and N.G. Leet, J.Food Sci., 1966, 31, 450.

39. C.L. Davey and K.V. Gilbert, Meat Sci., 1977, 1, 49.

40. C.L. Davey and M.R. Dickson, J.Food Sci., 1970, 35, 56.

41. C.L. Davey and K.V. Gilbert, J.Food Technol., 1967, 2, 57.

42. C.L. Davey and A.E. Graafhuis, J.Sci.Food Agric., 1976, 27, 301.

43. C.L. Davey and K.V. Gilbert, J.Food Sci., 1969, 34, 69.

44. R.H. Locker and D.J.C. Wild, Meat Sci., 1982, 7, 93.

45. O.A. Young, A.E. Graafhuis and C.L. Davey, Meat Sci., 1980, 5, 41.

46. C.L. Davey and K.V. Gilbert, J.Sci.Food Agric., 1974, 25, 931.

47. C.L. Davey, H. Kuttel and K.V. Gilbert, J.Food Technol., 1967, 2, 53.

48. A. Asghar and N.T.M. Yeates, CRC Crit.Rev.Food Sci.Nutr., 1978, 10, 115.

49. N.L. King and L. Kurth, "Fibrous Proteins: Scientific, Industrial and Medical Aspects", Academic Press, London, 1979, Vol.2, p.57.

50. R.H. Locker, "Advances in Meat Science", AVI Publ.Co., ⌐T, USA, Vol.1, 1983 (In press).

51. B.B. Marsh and W.A. Carse, J.Food Technol., 1974, 9, 129.

52. C.A. Voyle, J.Food Technol., 1969, 4, 275.

53. R.H. Locker and D.J.C. Wild, Meat Sci., 1982, 7, 189.

54. L.A. Staehelin, Int.Rev.Cytol., 1974, 39, 191.

55. R.O. Hynes and K.M. Yamada, J.Cell Biol., 1982, 95, 369.

56. C.L. Davey and K.V. Gilbert, J.Sci.Food Agric., 1975.

26, 755.

57. R.H. Locker and G.J. Daines, J.Sci.Food Agric., 1976, 27, 193.

58. B.B. Marsh, J.V. Lochner, G. Takahashi and D.D. Kragness, Meat Sci., 1980, 5, 479.

2

The Chemistry of Intramolecular Collagen

By Allen J. Bailey

A.R.C. MEAT RESEARCH INSTITUTE, LANGFORD, BRISTOL BS18 7DY, U.K.

INTRODUCTION

The fragile contractile elements of muscle fibres described in the previous chapter are supported and protected from damage and over-extension by layers of collagen fibres. These fibres provide a supporting framework within each muscle and separate different muscles, but their most important role is the efficient transmission of the contractile force of the muscular elements via the tendons to the skeleton, thereby producing movement. The ability of the intramuscular collagen and the tendons to transmit the contractile force with little loss of energy is dependent on the virtually inelastic properties of the collagen fibres. Collagen is the only component of muscle that possesses any substantial degree of mechanical strength, and has therefore been implicated in the toughness of meat. However, determining the precise mechanism of its role in texture is proving a complex problem.

Classical histological studies have identified collagen at various levels in the structure of muscle.[1] The muscle is surrounded by a fascia of thick collagen fibres extending from the tendons and this is referred to as the epimysium. Within the muscle the bundles of fibres are also surrounded by a sheath of collagen fibres referred to as the perimysium. These are finer fibres than in the epimysium and are often arranged in a laminated fashion around the bundles; they are sometimes called reticular fibres owing to their ability to take up silver stain. The muscle fibres themselves are surrounded by a continuous sheath known as the endomysium (Fig.1). On examination at high magnification in the electron microscope, the endomysium appears to be made up of an apparently amorphous basement membrane surrounding the myofibres, and an outer layer of very fine collagen fibres. These fibres not only surround the basement membrane but interconnect the myofibres (Fig.1b).

In order to determine the role played by these morphologically distinct collagenous tissues in both the functioning of the live muscle and in the texture of cooked meat, it is necessary to understand the molecular structure of the collagen molecule and the polymerised collagen fibre, and their interaction with associated proteins and proteoglycans in the

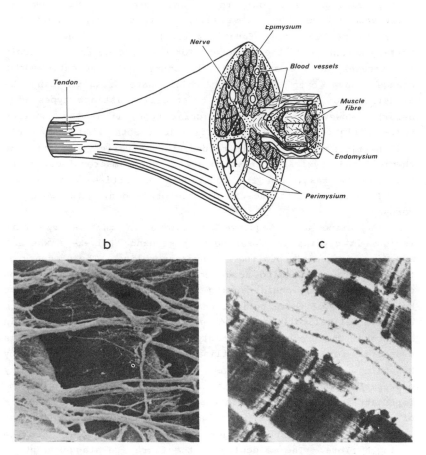

Figure 1 (a) Diagrammatic representation of the organisation of the epimysium, perimysium and endomysium in muscle; (b) Scanning electron micrograph of muscle fibres showing the inter-connecting collagen fibres; (c) Transmission electron micrograph of the endomysium showing the amorphous basement membrane and fine collagen fibres. (Fig.1b courtesy of Borg and Caulfield).

extracellular matrix. These interactions result in the widely
different morphological arrangements of connective tissue and
consequently lead to different functions of the collagen fibres
in various tissues.

MOLECULAR STRUCTURE OF COLLAGEN

The basic structure of collagen has been known for over
twenty years but it is only in the last few years that chemical
differences have been observed[2,3,4] that account for the
diversity of biological functions exhibited by collagen. The
primary structure of the different forms of the collagen molecule
is extremely simple and is highly conserved. Recently minor
changes have been identified which have resulted in the
classification of collagen into genetically distinct types (see
below). However, within a particular type, it is the numerous
post-translational chemical modifications both of the molecule
and the fibre that permit it to undertake a variety of different
chemical and physical interactions with other macromolecules and
result in a change in its functional properties; for example
Type I collagen in tissues as diverse as tendon, skin, bone and
cornea.

The amino acid sequence is unique in that every third
residue is glycine in the repeating sequence (Gly-X-Y) where X
and Y are frequently proline and hydroxyproline. The presence of
these heterocyclic imino acids confers a rotational restriction
on the polypeptide chain causing it to adopt a polyproline type
helix. Three of these helical chains are wound together to form
a triple helical collagen molecule 300 nm in length and 1.5 nm in
diameter. The presence of the glycine at every third residue is
essential for internal stabilisation of the triple-helix by
H-bonding. The side chains consequently project radially
outwards from the axis of the molecule and therefore become
important in intermolecular interactions. These interactions are
of particular importance in alignment of the molecules in the
well established quarter-stagger end-overlap fashion to form the
collagen fibre. The molecules in the fibre are staggered by 67
nm (D), the axial periodicity observed in the electron microscope, where 4.4 D represents the length of the molecule. The
precise orientation of the molecules is governed by the
asymmetric distribution of the charged residues along the length

of the molecule whilst the hydrophobic groups play a role in maintaining the organisation of the fibre.

At this stage the aggregated collagen monomers do not possess any mechanical strength, but polymerisation through the formation of covalent intermolecular cross-links occurs as the fibre grows thereby providing tensile strength.[5,6] The ultimate strength of the particular tissue depends, therefore, not only on the extent and nature of these cross-links, but also on the orientation of the fibres within the particular tissue.

GENETICALLY DISTINCT TYPES OF COLLAGEN

Over the past few years detailed chemical analysis has established the presence of several genetically distinct collagens which are, to a certain extent, tissue specific.[7,8] Among the fibrillar collagens Type I is the major collagen of skin, tendon and bone, Type II of cartilage and intervertebral disc, Type III of foetal skin and the vascular system. Type IV collagen is non-fibrillar and is the major component of amorphous basement membrane. Type V is closely associated with basement membrane but appears to consist of fine fibres, although this has not been confirmed. The Type M group of pericellular cartilage collagens[9] are probably also non-fibrillar but have not yet been completely characterised (Table 1).

Biochemical and immunochemical techniques have established

Table 1   Genetic types of collagen in muscle

	Molecular configuration	Characteristics	Major location
I	$[\alpha 1(I)]_2 \alpha 2$	thick fibres	epimysium
III	$[\alpha 1III]_3$	disulphide bond thin fibres	perimysium
IV	$\alpha 1(IV)$ $\alpha 2(IV)$	non-fibrillar disulphide bond	endomysium
V	$[\alpha IV]_2 \alpha 2V$	fine fibres	endomysium

that in muscle the fascia or epimysium is composed of Type I
collagen, the perimysium is mainly Type III collagen with varying
amounts of Type I, and the endomysium contains both Type IV and
Type V collagen[10] (Fig.2). The tendon at the end of the muscle
is, like the epimysium, mainly Type I collagen, although imuno-
fluorescence studies demonstrate that the large bundles of Type I
fibres within the tendon are surrounded by a fine sheath of Type
III and Type IV collagen.[10] All the fibrous collagen types show
the typical axial periodicity indicating a quarter-stagger
alignment and they are all cross-linked by the same enzymic
mechanism at the same location in the molecule.[11] Recent studies
have shown that the non-fibrillar collagens (Types IV and M) are
also stabilised by the same types of cross-links[12] but the
precise intermolecular location of the cross-link has not yet
been elucidated. Preliminary evidence on Type IV collagen
indicates that the cross-links are also initiated from the N- and
C-terminal ends of the molecule[13,14] despite the completely
different molecular organisation.

INTERMOLECULAR CROSS-LINKING

Since the nature and extent of the cross-links determine the
tensile strength of the collagen fibre, considerable effort has
been made to elucidate their structure, location and change with
age. It is now generally accepted that the cross-links are
formed by oxidative deamination of specific lysine or hydroxylysine residues in the short non-helical regions at both ends of
the molecule.[4] The enzyme involved, lysyl oxidase, acts on the
fibre and converts the εNH_2 group of the lysine to an aldehyde.[15]
The precise alignment of the molecules ensures that the aldehyde
group is located adjacent to a specific hydroxylysine (residue
945) within a highly conserved sequence (X-hyl, gly, his, arg) in
all three fibrous collagens. Similar reactions occur through the
lysine aldehydes at the C-terminus of the molecule, reacting with
a hydroxylysine (residue 103) near the N-terminus of the triple
helix and again within a similar amino acid sequence (Fig.3).
This specific environment presumably initially ensures location
of the enzyme on the triple helical part of the molecule close to
the lysine residue in the N-terminal region of an adjacent
molecule, and subsequently enhances the reaction of the aldehyde
formed with the εNH_2 group of the hydroxylysine in the helix to

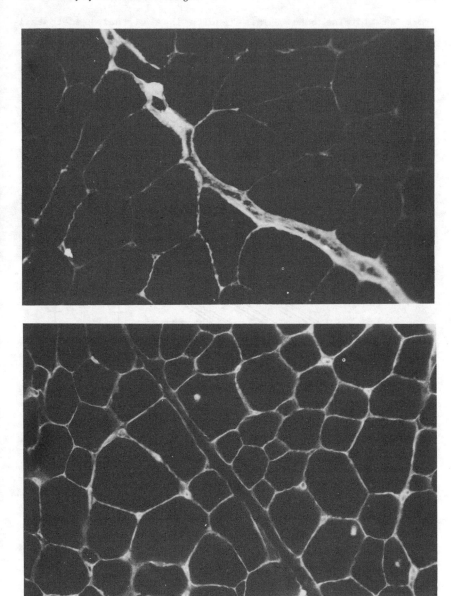

Figure 2 Immunofluorescence staining of muscle cross-sections showing (upper) the location of Type III collagen in the perimysium and (lower) Type IV in the endomysium.

form an aldimine bond.   This results in a virtual head-to-tail
polymerisation of the molecules (Fig.3).

Stabilisation of the fibres with an aldimine bond derived
from lysine-aldehyde results in a cross-link labile to acid
solution and temperatures over 65°, for example in young dermal
collagen and tendon.   Where hydroxylysine aldehyde is the
cross-link   precursor,   the   initial   aldimine   cross-link
spontaneously undergoes an Amadori rearrangement to form a stable
keto-imine   cross-link.   Tissues possessing this keto-imine
cross-link   are   insoluble   in   acid   solution   and   on   thermal
denaturation, for example bone and cartilage.   Both types of
cross-link are of course stable under physiological conditions

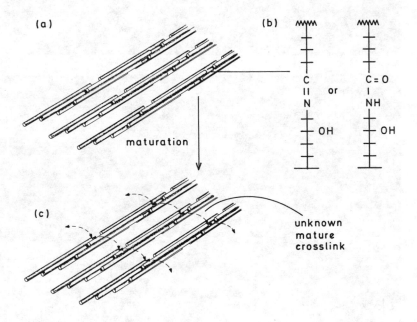

Figure 3   Location of the covalent intermolecular cross-links;
(a) head-to-tail in the immature collagen microfibril; (b) nature
of cross-links in immature collagen, the labile aldimine bond and
the stable keto-imine bond; (c) proposed location of the inter-
microfibrillar cross-links in mature collagen.

and the head-to-tail longitudinal cross-linking builds up a long microfibril and confers considerable tensile strength, but to prevent slippage between the microfibrils interfibrillar (transverse) cross-links are necessary. The increase in tensile strength of the fibre with increasing age is believed to be due to an increase in these interfibrillar or transverse crosslinks.[6] The chemistry of these bonds has not yet been elucidated but recent evidence suggests that the transverse bonds are derived from the longitudinal lysine-aldehyde cross-links, and indeed these latter cross-links decrease with maturation concomitant with an increase in strength of the tissue. Unlike the aldimine and keto-imine cross-links, the interfibrillar and 'mature' cross-links are not reducible and it has therefore proved difficult to characterise them.

Cyanogen bromide (CB) cleavage of mature collagen has revealed the presence of a high molecular weight cross-linked CB peptide, and detailed analysis has revealed the component to be composed primarily of αICB6 and αICB5 indicating a polymeric peptide CB6-CB5-CB6-CB5- etc. It is highly probable that a similar polymer is also formed from the N-terminal peptide, i.e. (CB6-CB1)n but the small size of the CB1 peptide prevents its detection in a high molecular weight polymer. These results indicate that in mature tissues the transverse cross-linking occurs between molecules in register.[16] This type of cross-linking could be achieved on the quarter-staggered microfibril model only by intermicrofibril cross-linking provided the microfibrils were in register to allow cross-linking between the end-terminals. Similarly the alternative quasi-hexagonal model can accommodate these tail-to-tail links in certain configurations.[17] We have proposed[6,16] that in this way a network of longitudinal and transverse cross-links can be generated to stabilise the fibres involving no other reactions than those initiating from the N- and C-terminal lysine-aldehydes. Furthermore, the cross-linking can continue spontaneously without the difficulty of involving further enzyme reactions within the fibre (Fig.3b).

Despite various reports, no convincing evidence has been provided for interhelical cross-links. Several stable non-reducible cross-links derived from the precursor lysine-aldehyde have been described; the structure currently demanding considerable attention is the pyridinoline cross-link proposed by Fujimoto et al.[18] The location of pyridinoline as a cross-link between identified CNBr peptides, which would confirm its proposed role as the cross-link stabilising mature collagen, has not been forthcoming. In contrast, the polymeric αICB6-5, derived from mature collagen, which by definition should contain the 'mature' cross-link does not contain pyridinoline.[16] It is possible that pyridinoline fulfils another role. More recently Fujimoto[19] has proposed that histidinoalanine is the 'mature' cross-link since it increases with age in contrast to pyridinoline which decreases after maturity.   However, histidinoalanine does not appear to be specific for collagen, being detected in several other proteins.

THERMAL DENATURATION

The collagen molecule is highly crystalline and possesses a fairly sharp melting point, collapse of the triple helix occurring around 39° for mammalian collagen.   The precise alignment of the molecules in the fibre increases the energy of crystallisation, and the fibre denatures at about 65°.[20]   On denaturation from the long rigid molecules of collagen to random chains of gelatin, the fibre collapses if unrestrained to a swollen elastic fibre one quarter of its length resulting in a dramatic ten-fold decrease in tensile strength.   The residual tensile strength of this elastic fibre, and indeed the tension generated during contraction, is determined by the proportion of thermally stable interchain cross-links.   Native fibres stabilised by the aldimine cross-link (dehydro-hydroxylysino-norleucine) will exhibit little strength following denaturation since these cross-links are thermally labile.   Continued heating would indeed rupture all the cross-links and render the fibre soluble.   In contrast, fibres stabilised by the keto-imine cross-link (hydroxylysino-5-keto-norleucine) maintain some strength since this cross-link is thermally stable.   This difference can be most clearly demonstrated by denaturation under isometric conditions and determination of the tension generated during thermal denaturation.   Allain et al.[21] showed a direct correlation between the extent and permanence of the tension generated and the nature of the cross-link.   Foetal rat skin, which contains a high proportion of the thermally stable keto-imine cross-links, generates a greater tension than skin from a 3-week old animal where the major stabilising cross-link is the thermally labile aldimine bond. More important are the observations of rapidly increasing tension with age and the ability in old animals to maintain a higher proportion of the maximum tension even at high temperatures (Fig.4). These latter observations result from an increasing replacement of the labile aldimine bonds by thermally stable 'mature' cross-links.

Similar isometric tension studies have been carried out on epimysium,[22] which like tendon is primarily Type I collagen and possesses a high proportion of the aldimine cross-link in addition to the keto-imine cross-link. The perimysium exhibits a similar effect but the results are more difficult to analyse since this tissue contains both Type I and Type III collagen.[10] The high proportion of the keto-imine cross-link also makes this tissue somewhat different from the epimysium.[23] Few studies have been carried out on the thermal stability of basement membrane

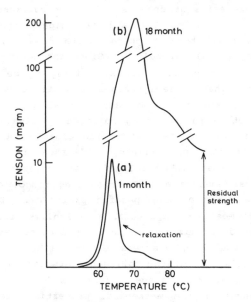

Figure 4 Isometric tension studies showing: (a) tension generated and rapid relaxation of immature tendon due to rupture of thermally labile bonds on further heating after contraction; (b) markedly increased tension generated and high proportion of residual tension at high temperatures due to thermally stable cross-links in mature tendon.

collagen. Although the Type IV molecule appears to possess a
higher T_D than the Type I collagen molecule, 41° and 39°
respectively,[24] the intact basement membrane melts and shrinks
over a wide range. Values over the range 55° to 75° have been
recorded for both lens capsule basement membrane and for
extracted endomysial sheath.[25] This lower initial melting
temperature and wide range is presumably due to the lower energy
of crystallisation in the non-fibrillar organisation of the
membrane and the high proportion of proteoglycan, preventing the
same extent of interaction between the individual collagen
molecules. The basement membrane is stabilised by the thermally
stable keto-imine cross-link, but the tension generated on
thermal denaturation of the membrane has not been determined,
although it is likely to be much less than fibrillar collagen.

ACTION OF PROTEOLYTIC ENZYMES

The collagen fibre is generally believed to be degraded in
vivo by a complex series of enzymes.[4,26] The collagenases have
an optimum activity at pH 7.4 and are highly specific for
collagen, cleaving all three peptide chains at the same locus
along the triple helical molecule.[27] Since the optimum activity
of collagenases is at pH 7.4, it is unlikely to be active in
meat. On the other hand, the lysosomal catheptic enzymes have a
pH optimum of 3.5 and have some activity at pH 5.5. These
enzymes are effective in degrading collagen by cleavage of the
peptide chains in the non-helical region close to the lysine-
derived intermolecular cross-links,[28,29] thus releasing the
molecule from the cross-linked polymer. At temperatures around
37° the fragments denature and the resulting random chains are
susceptible to a range of non-specific proteases. During the
conditioning of meat no appreciable solubilisation of the
collagen could be detected. However, limited proteolysis of the
collagen could result in weakness of the fibre without any
detectable solubilisation of the collagen. Studies in vitro have
shown that very limited cleavage by cathepsins can be readily
detected by changes in the mechanical properties, for example
with the more sensitive isometric tension measurements.[30]

Proteases specific for the degradation of basement membrane
collagen have been identified in pathological situations, but
their presence in meat has not been demonstrated. The loss of

integrity of the endomysium basement membrane could significantly affect the properties of the meat, particularly its water-holding capacity.

ROLE OF COLLAGEN IN THE TEXTURE OF MEAT

It is these basic studies on the structure of collagen that have permitted a rational description of its role in the texture of meat. The proportion of collagen in muscle is small, generally about 2%, although a few muscles contain up to 10%.[31] The latter produce tough meat suggesting a direct relationship between collagen content and texture, but in reality such a relationship could not be established.[32] The difficulties in the past included explaining the difference in texture between different muscles from the same animal and of identical muscles obtained from calf and adult cattle. There is little change in the amounts of collagen with age, but the collagen in calf muscles may be exuded from the cooked meat and set as gelatin on cooling, whilst the collagen of the adult muscle is insoluble and is retained within the meat. This difference can now be explained by the change in both the nature and extent of the cross-links with increasing age.

As the temperature of a piece of meat is raised to between 40° and 50°, there is an increase in toughness as determined by the shear value, and this has been interpreted as being due to denaturation of the myofibrillar proteins, primarily the actomyosin complex[33,34] (Fig.5). The solubility of the myofibrillar proteins decreases dramatically, probably due to the coagulation of unfolded peptide chains, but this insolubility is apparently not due to the formation of disulphide bonds. Denaturation results in a shrinkage of the muscle fibre as has been demonstrated by a number of workers[35,36] and most recently in a more detailed study by Bendall and Restall.[36] According to the latter, the shrinkage occurs totally in the transverse direction, the muscle fibre shrinking within the endomysium with a consequent loss of water. In a subsequent chapter Offer[37] proposes that the loss of water occurs through the resulting spaces between the endomysium and the shrunken fibre under pressure from the stressed native collagen network (Fig.6). It should be remembered that at this temperature the collagen is unaffected.

As the temperature is raised further there is only a slow increase in shear value, but at 60-70° there is a second rapid increase and a further loss of fluid from the meat (Fig.5). This second shrinkage, in contrast to the first, has been reported to be almost entirely longitudinal[35,36] but it is unlikely that contraction of the perimysium does not involve transverse shrinkage. Indeed, in view of the laminated cross-ply arrangement of the fibres, contraction would be expected to result in shrinkage in all directions. The extent of this rapid increase is dependent on the age of the animal, and is concomitant with the observed gelatinisation of the collagen of the epi- and perimysium at about 65°.

As discussed above, on denaturation at 60-70° collagen shrinks to a quarter of its original length, and its residual elastic strength depends on the extent and nature of the intermolecular cross-links. If contraction is restrained then a tension is generated, the extent of which is again dependent on the number and nature of the covalent cross-linking. On considering the contraction of collagen in meat we must therefore consider both the extent of the tension generated against the muscle during shrinkage of the collagenous sheath and the residual strength of the now denatured fibres interconnecting the endomysium of the muscle fibres.

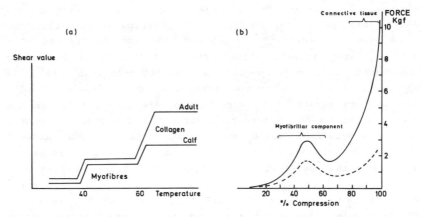

Figure 5 (a) Diagrammatic representation of the changes in the shear value of meat with increasing temperature for meat from a young and old animal; (b) Force-compression curve as determined using an Instron for meat from a 3-month old calf and a 7-year old cow.

We have previously proposed[38] that the tension generated during shrinkage of the perimysial collagen at 60-70° results in a compression of the meat in which the muscle fibres would be squeezed together. These muscle fibres contain denatured myofibrils and water (Fig.6b) and compression would involve loss of water and hence closer contact of the muscle fibres within a fibre bundle. The endomysium also denatures at this temperature but the extent of this contraction is unknown at present. If mild it could increase the loss of water from the muscle fibre but if considerable could result in compression of the myofibrils within the fibre. This dramatic loss of fluid following compression presumably occurs through the channels between the endomysium and the denatured muscle fibre as described by Offer,[37] but possibly also via the now widened channels in the denatured perimysium. At even higher temperatures further shrinkage of the denatured fibres themselves might lead to increased protein-protein interaction due to dehydration of the myofibrils. Contraction of the collagen readily accounts for the excess fluid expelled from the meat, but this would not necessarily result in an increase in the toughness of the meat. A compression of the fibre bundles is thought to be necessary to account for the increase in toughness by increasing the protein

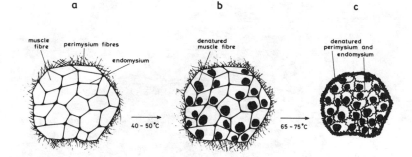

Figure 6 Effect of heat on muscle proteins; (a) cross-section of rigor muscle; (b) cross-section of muscle heated to 45-50° showing shrinkage of the muscle fibre proteins proteins on denaturation; at this stage the collagen of the peri- and the endomysium is unaffected; (c) proposed compression of the muscle fibres after thermal shrinkage of the peri- and endomysium at 65-70°C.

content per unit area in shear value determinations. Further, since the tension generated increases with age, the resulting increased compression would account for the increased toughness of the meat with age of the animal.

In addition, the residual strength of the denatured collagen maintaining the adhesion between the muscle fibre bundles must also play a part in determining the toughness of the meat. The binding together of the muscle fibre bundles can be observed on gently teasing apart the cooked fibres, when the denatured collagen can be extended until it ruptures, the denatured myofibre bundle remaining intact. These strands are clearly the weak point in the structure and therefore determine the ultimate toughness of the meat (Fig.7). During cooking there is therefore a reversal in the relative mechanical properties of the muscle fibres and collagen. The denatured muscle fibres aggregate to form a more rigid gel whilst the denatured collagen possesses a fraction of its original strength. The tensile strength of the now gelatinised elastic fibre is dependent on the age of the animal. Clearly the more mature the animal the greater the residual strength, resulting in greater adhesion between the denatured muscle fibres and hence a tougher meat.

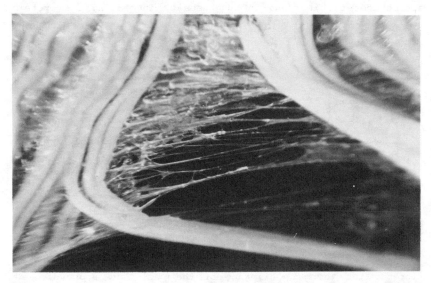

Figure 7 Denatured collagen fibres being extended by teasing apart the denatured muscle fibres.

   Thus it is our contention that the extent of the compression due to collagen shrinkage, together with the residual strength of denatured collagen, can account for the sudden change in the toughness at 60-70° as demonstrated by the shear value, and for the increased exudation of fluid. The relationship of the extent of the increase in shear value at 60-70° with increasing age of the animal, and the known increase in thermal stability of the collagen with age, are further support for the proposal that this change is primarily due to the contraction of collagen.

   It is unlikely that there would be such a dramatic change at 65° in the already denatured actomyosin, although some workers have reported a decrease in the sarcomere length.[33,35] This contraction could result in an increase in toughness similar to that observed with cold-shortened muscle, but it is very unlikely that this effect could be age-related in the way that the changes in collagen correlate with the increase in toughness or result in a rapid expulsion of fluid.

   During conditioning the myofibrils are clearly affected as demonstrated by the disappearance of the Z-bands,[39,40] but the effect on collagen is less easily discerned. However, the reduction in tension generated during thermal contraction[30] can be ascribed to the limited cleavage of the peptide bonds in the non-helical cross-linking region. This effect combined with the lower residual strength of the denatured fibre would certainly reduce the toughness of the collagen, thus indicating a role for collagen in the tenderisation of meat during conditioning. Although the mechanism of action of the proteases on the myofibrils and the collagen are, at least partially, understood, the relative contribution of each has not yet been determined.

   Although there is now a clear rationale for the role of collagen in meat texture, the precise details have not yet been worked out. The endomysium is an amorphous basement membrane and, like the more classical lens capsule basement membrane, has a wide-ranging shrinkage temperature, 60-75°, but for technical reasons its strength and hence its effect on the myofibrils during denaturation have not been determined. In the initial stages of heating the myofibres shrink away from the endomysium, and the non-fibrous nature of the endomysium suggests that it might not exert much effect on the myofibres during denaturation. In some muscles the endomysium is associated with a higher proportion of fine collagen fibres, but the effect of the denaturation of these fibres is difficult to evaluate. However, the perimysium is composed of thicker fibres of Type III and Type I collagen which on denaturation would certainly exert a tension leading to compression of the muscle fibre bundles and loss of water.

The organisation of the perimysium in muscle may also affect the degree of tension exerted on the myofibres. In certain muscles the perimysium is laminated,[41] each layer being at an angle to the next so that on denaturation the tension generated would be different from that exerted by fibres which were all aligned uniaxially as in tendon. Further studies on the structure and properties of the basement membrane of the endomysium are clearly required to determine the relative effects of the peri- and endomysium on texture.

As discussed above, the extent of the tension generated and the residual strength of the fibres are dependent on the nature of the cross-linking. The perimysium has already been shown to possess a higher proportion of the thermally stable keto-imine cross-link than the epimysium of the same muscle.[38] It may also possibly depend on the ratio of Type I to Type III collagen present in the perimysium although our recent studies suggest little difference between muscles. At the present time fibres composed solely of Type III fibres have not been identified, hence the effect of the disulphide bonding at the C-terminal ends of these molecules on the tension generated during shrinkage cannot be determined. Although it is clear that both tension and residual strength are important, the relative importance of each on the increase in toughness or shear value remains to be elucidated.

In the final analysis by the consumer the sensation of toughness is provided by the major proteins in meat, i.e. actin and myosin, but the collagen, although only constituting 2% of the protein, plays an important role in determining the quality of that sensation. The extent of the role played by the collagen is determined partly by the total amount of intramuscular collagen but primarily by the quality of the collagen in terms of its stabilising cross-links.

REFERENCES

1. W. Bloom and D.W. Fawcett, "A Textbook of Histology", W.B. Saunders, Philadel., USA (10th Edn), 1975.

2. G.N. Ramachandran and A.H. Reddi, "Biochemistry of Collagen", Plenum Press, New York, 1976.

3. P. Bornstein and W. Traub, "The Proteins", Academic Press, New York, 1979, Vol.4.

4. A.J. Bailey and D.J. Etherington, "Comprehensive Biochemistry", Elsevier Publ.Co., Amsterdam, 1980, Vol.19B, p.299.

5. A.J. Bailey, S.P. Robins and G. Balian, Nature, 1974, 251, 105.

6. N.D. Light and A.J. Bailey, "Fibrous Proteins: Scientific, Industrial and Medical Aspects", Academic Press, London, 1979, Vol.1, p.151.

7. P. Bornstein and H. Sage, Ann.Rev.Biochem., 1980, 49, 957.

8. K. Kuhn, "Immunochemistry of the Extracellular Matrix," CRC Press Inc, Florida, USA, 1982, Vol.I, p.1.

9. M. Shimokomaki, V.C. Duance and A.J. Bailey, FEBS Lett., 1980, 121, 51.

10. V.C. Duance, D.J. Restall, H. Beard, F.J. Bourne and A.J. Bailey, FEBS Lett., 1977, 79, 248.

11. P.P. Fietzek, H. Allmann, J. Ranterberg and G. Wachter, Proc.Nat.Acad.Sci.(USA), 1977, 74, 84.

12. M. Shimokomaki, Ph.D. Thesis, University of Bristol, 1982.

13. H. Dieringer and R.W. Glanville, Coll.Rel.Res., 1983, 3, 65.

14. A.J. Bailey, T.J. Sims and N.D. Light (In press).

15. R.C. Siegel, Int.Rev.Conn.Tiss.Res., 1979, 8, 73.

16. N.D. Light and A.J. Bailey, Biochem.J., 1980, 185, 373.

17. A.J. Bailey, N.D. Light and E.D.T. Atkins, Nature, 1980, 288, 408.

18. D. Fujimoto and T. Moriguchi, J.Biochem.(Tokyo), 1978, 83, 863.

19. D. Fujimoto, M. Hirama and T. Iwashita, Biochem.Biophys. Res.Comm., 1982, 104, 1102.

20. W.F. Harrington and P.H. von Hippel, Adv.Protein.Chem., 1961, 16, 1.

21. J.C. Allain, M. Le Lous, S. Bazin, A.J. Bailey and A. Delauney, Biochim.Biophys.Acta, 1978, 533, 147.

22. J. Kopp and C. Valin, Ann.Technol.agric., 1979, 28, 107.

23. A.J. Bailey and T.J. Sims, J.Sci.Food Agric., 1977, 28, 565.

24. R.A. Gilman, J. Blackwell, N.A. Kefalides and E. Tomichek, Biochim.Biophys.Acta, 1976, 427, 492.

25. A.J. Bailey, unpublished data.

26. D.J. Etherington, "Protein Degradation in Health and Disease". Ciba Found.Symp.75, 1980, p.87.

27. D.E. Woolley and J.M. Evanson, "Collagenase in Normal and Pathological Connective Tissue", John Wiley, Chichester, 1980.

28. A.J. Barrett, "Proteinases in Mammalian Cells and Tissues", North Holland Biomed. Press, 1977.

29. D.J. Etherington and P.J. Evans, Acta biol.med.germ., 1977, 36, 1555.

30. J. Kopp and C. Valin, Meat Sci., 1981, 5, 319.

31. J.R. Bendall, J.Sci.Food Agric., 1967, 18, 553.

32. A.J. Bailey, J.Sci.Food Agric., 1972, 23, 995.

33. J.G. Schmidt and F.C. Parrish, J.Food Sci., 1971, 36, 110.

34. R.L. Hostetler and W.A. Landmann, J.Food Sci., 1968, 33, 468.

35. J.F. Aronson, J.Cell Biol., 1966, 30, 453.

36. J.R. Bendall and D.J. Restall, Meat Sci., 1983, 8, 93.

37. G.W. Offer. Chapter 5.

38. T.J. Sims and A.J. Bailey, "Developments in Meat Science - 2", Appl.Sci.Publ., 1981, p.29.

39. W.R. Dayton, D.E. Goll, M.H. Stromer, W.J. Reville, M.G. Zeece and R.M. Robson, "Proteases and Biological Control", Cold Spring Harbor Conferences on Cell Proliferation, 1975, Vol.2, p.551.

40. E. Dransfield and D.J. Etherington, "Enzymes and Food Processing", Applied Science Publ.Ltd, London, 1981, p.177.

41. R.W.D. Rowe, J.Food Technol., 1974, 9, 501.

3
The Control of Post-Mortem Metabolism and the Onset of *Rigor Mortis*

By Robin E. Jeacocke

A.R.C. MEAT RESEARCH INSTITUTE, LANGFORD, BRISTOL BSI8 7DY, U.K.

INTRODUCTION

It is generally the case, in the meat we consume, that ATP has long since disappeared and, as a consequence, the ATP-dependent processes which characterise the living muscle cell have ceased to take place. Many features of meat which are important from the consumer's point of view are determined by the manner in which this ATP-depleted state is attained and it is therefore highly desirable to attempt to understand the metabolic processes which occur as ATP-replete muscle is transformed into ATP-free meat. It is important to recall that the type of chemical and associated mechanical changes which occur in muscle post mortem are the outcome of those particular characteristics of skeletal muscle which fit this tissue for its biological role as a contractile machine. In particular, the cellular control mechanisms which modulate power output during the normal activities of the muscle fibre are of central importance in setting the rate at which events proceed post mortem, whilst the stiff and inextensible state which muscle fibres finally achieve in rigor mortis results from interactions somewhat similar to those which cause muscle contraction in the living cell. Some appreciation of the manner in which a muscle cell normally functions is therefore necessary in order to understand the abnormal but (from the viewpoint of the meat industry) crucial events which take place as the process of ATP resynthesis finally fails within each cell and the muscle is converted into meat.

I will endeavour to outline the metabolic changes which take place post mortem, to show how the progress of these events is controlled and also to indicate how alterations in the rate at which these events proceed may affect meat quality. Finally, I shall attempt to show how some recent experimental work clarifies certain features of these post-mortem changes which were previously obscure.

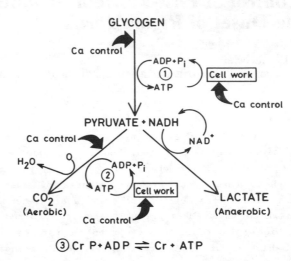

Figure 1 Aerobic and anaerobic metabolism of muscle glycogen.
ATP resynthesis from ADP plus inorganic phosphate (P_i) takes
place by substrate level phosphorylaton (1) and by oxidative
phosphorylation (2). In addition, ATP resynthesis can occur (3)
at the expense of the conversion of creatine phosphate (CrP) to
creatine (Cr). This latter reaction is catalysed by the enzyme
creatine kinase and is near equilibrium in the resting cell. The
points at which Ca^{2+} ions exert control over the metabolic
pathways are indicated.

METABOLIC AND MECHANICAL CHANGES POST MORTEM

When an animal is killed, the heart stops beating and as
a result the circulatory system ceases to supply the muscles with
metabolisable fuels such as glucose and with oxygen. In order to
replenish the ATP which is being continually hydrolysed to power
the various energy-consuming activities which constitute cellular
work (such as ion transport and muscle contraction) skeletal
muscle cells have subsequently to rely upon the anaerobic
metabolism of their intracellular fuel.

The storage carbohydrate glycogen (Fig.1) is the major fuel
for ATP resynthesis under these post-mortem conditions. The
breakdown of glycogen to glucose 1-phosphate (glycogenolysis) and
its subsequent conversion via the glycolytic pathway to pyruvate
and reducing equivalents (NADH) is accompanied by the resynthesis
of a small quantity of ATP (3 ATP per hexose unit of glycogen) in

so-called substrate-level phosphorylation. In the aerobic cell, oxidation of these reducing equivalents and of pyruvate occurs within the mitochondria to produce carbon dioxide and water with the resynthesis (by oxidative phosphorylation) of a much larger amount of ATP (36 ATP per hexose unit of glycogen). This large capacity for substrate oxidation and ATP resynthesis is not available to the anaerobic cell, which is forced to reoxidise the NADH produced in glycolysis by catalysing the reduction to lactate of the pyruvate which is produced simultaneously. An acceptor (NAD+) for further reducing equivalents is thereby regenerated and lactate is thus the end product of anaerobic glycolysis.

One additional source of ATP resynthesis exists within the anaerobic muscle cell; this is the reservoir of creatine phosphate, the phosphate groups of which can be donated to ADP to produce creatine and ATP. This reaction, catalysed by the enzyme creatine kinase, occurs at very high rates in muscle fibres. The intracellular activity of creatine kinase is so high that the reaction is practically at equilibrium in a resting muscle cell. This fact, long suspected but rather difficult to demonstrate, has recently been substantiated by elegant kinetic studies using phosphorus NMR.[1]

Despite the rather small yield of ATP which can be obtained from anaerobic glycolysis, the glycogen reserves of skeletal muscle cells (normally equivalent to more than 100 mM ATP) can underwrite anaerobic ATP resynthesis for several hours in a resting muscle cell. The creatine phosphate reserves, by contrast, are smaller, typically equivalent to 20-30 mM ATP.[2]

The muscle cell is, of course, functionally highly specialised to perform mechanical work in a well controlled manner. The maximum power output (and maximum rate of ATP hydrolysis) exhibited by skeletal muscle is, according to the standards of other cells, remarkably high, whilst the minimum rate of energy expenditure and of ATP hydrolysis in the relaxed, resting cell is several hundred-fold lower than this. Despite such large variations in the rate of ATP breakdown, the intracellular concentration of ATP remains essentially constant at about 5 mM. This implies that the rate of ATP resynthesis must always match, rather precisely, the rate of ATP breakdown. This match or compensation of rates is achieved in two distinct ways. Firstly, the rate of ATP regeneration by glycolysis and, in an aerobic cell, the rate of oxidative phosphorylation alters to accommodate variations in the rate of ATP expenditure. A second method of compensation is available, however, and is part- icularly apparent at high rates of energy expenditure. The rate of anaerobic regeneration of ATP by glycolysis may be varied by perhaps 30- to 40-fold, but the highest rates of ATP hydrolysis in a contracting muscle cannot be equalled by glycolytic regeneration and, under these conditions, the creatine kinase catalysed regeneration of ATP from creatine phosphate is called into play, since this reaction alone is fast enough to cope. The equilibrium position of this reaction lies well over towards the production of creatine plus ATP and very little ADP is present in a near-equilibrium mixture of the reactants containing plenty of creatine phosphate (as in a normal relaxed muscle cell). The very high catalytic activity of the enzyme creatine kinase ensures that, even at very high rates of ADP production, the rate of ATP resynthesis via this reaction keeps pace and prevents the prevailing ATP concentration from falling appreciably. The reaction is, of course, no longer poised near equilibrium under these high flux conditions and the free ADP concentration rises substantially, to fall rapidly once more when the muscle relaxes and lower rates of ADP production again prevail.[1]

What causes the rate of ATP regeneration by glycolysis to alter in response to changes in the rate of ATP expenditure? It appears that this is the outcome of two distinct phenomena. Firstly, the progress of the glycolytic reactions which synthesise ATP requires a continued supply of ADP and the rate of this supply, which limits the rate at which glycolysis can proceed, depends upon the prevailing rate of ATP expenditure. Secondly, the activity of certain enzymes catalysing key steps in the sequence of reactions leading from glycogen to lactate may be modified by certain small molecules and ions, the intracellular concentrations of which alter as a result of alterations in the rate of cellular energy expenditure. The enzyme activity is enhanced when energy expenditure increases, and diminishes when the expenditure falls.

Various small molecules and ions act as effectors in this way, but a particularly important role is played by the calcium ion. Despite the fact that the external environment is rich in

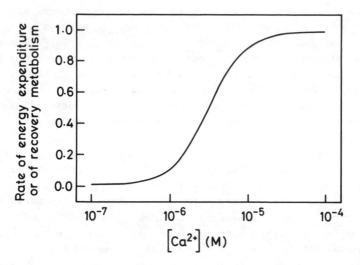

Figure 2 The calcium dependence of contractile energy expenditure and that of recovery metabolism. The precise shape of the observed dependence upon Ca^{2+} depends somewhat upon circumstances.

Ca^{2+} the concentration of Ca^{2+} free in the cytoplasm is held very low ($<10^{-7}$ M) during life as a result of the action of various energy-dependent Ca^{2+} pumps which are located in the plasma membrane at the cell surface and also inside the cell in the extensively ramifying sarcoplasmic reticulum membrane described by Dr Davey in Chapter 1. These various Ca^{2+} pumps translocate Ca^{2+} out of the cytoplasm in opposition to the steep inwardly-directed Ca^{2+} electrochemical gradient which otherwise would drive Ca^{2+} continuously into the cell cytoplasm. This energy-dependent Ca^{2+} translocation against the prevailing Ca^{2+} gradient is powered by the hydrolysis of ATP.

If the Ca^{2+} concentration free in the cytoplasm rises much above the normal resting level, various dramatic consequences ensue. As is now well known, the contractile apparatus is progressively activated by increases in Ca^{2+} ion concentration around 10^{-6} M and so the rate of cellular energy expenditure is enhanced in a Ca^{2+}-dependent fashion (Fig.2).

At the same time, however, the resynthesis of ATP both by glycolysis and aerobically by oxidative phosphorylation is enhanced by the Ca^{2+}-dependent activation of key enzymic steps located at the beginning of each of these pathways (Fig.1). Furthermore, the concentration-dependence of the activation of recovery metabolism (anaerobic and aerobic) by Ca^{2+} is indeed rather similar to that of contractile activation itself; consequently ATP expenditure and resynthesis are both activated by similar conditions.

The initial reaction (glycogenolysis) which converts glycogen into hexose phosphate units for further metabolism is now known to be Ca^{2+}-activated albeit in an indirect manner,[3] and for a summary see Cohen.[4] Likewise, certain key mitochondrial dehydrogenases involved in the oxidative phosphorylation pathway are similarly dependent upon Ca^{2+} ions for maximal catalytic activity[5] - see the scheme in Fig.2.

These various interacting metabolic processes and their associated control phenomena determine the manner in which events take place in muscle post mortem.

Fig.3 indicates the time course of the major post-mortem events in a mammalian muscle at a constant temperature of 30°C. The pH, which starts near neutrality, declines at an approximately constant rate as glycogen is converted into lactate with the production of one H^+ per lactate generated. Since the buffering power of the muscle remains practically constant across this pH range, the constant rate of pH fall reflects the fact that lactate production by glycolysis also proceeds at an almost constant rate. In the relaxed aerobic muscle just prior to death, as much as 80% of the total creatine may be phosphorylated as creatine phosphate, and in the anaerobic muscle post mortem it can be seen that this percentage declines as phosphocreatine is progressively lost to underwrite ATP resynthesis. The ATP content, by contrast, remains high since, as indicated earlier, a condition of near-equilibrium prevails in the creatine kinase catalysed reaction and the position of equilibrium lies well over towards creatine plus ATP. Anaerobic glycolysis, under these conditions, unlike oxidative phosphorylation in the aerobic cell prior to the death of the animal, is evidently unable to provide an ATP resynthetic pressure which is high enough to uphold the high creatine phosphate content characteristic of the relaxed

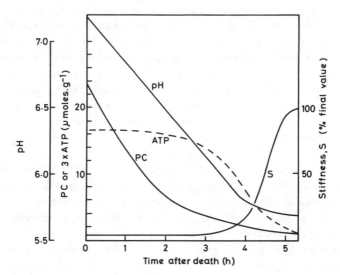

Figure 3 Time course of the main chemical and mechanical events post mortem in a beef sternomandibularis muscle held at 37°C. PC-phosphocreatine.

muscle fibre in the normally functioning animal – see Jeacocke[6] for some probable energetic consequences of this fact.

Subsequently, when the creatine phosphate reserves are largely exhausted, the ATP content also begins to decline. The glycogen reserves of the muscle fibre are able to continue to support ATP resynthesis, but they too are finite and ultimately become depleted, by which time lactate production has lowered the muscle cell pH to below 6 in an originally glycogen-replete cell. Finally, the stiffness of the preparation begins to rise, signalling the onset of rigor mortis as described by Dr Davey (Chapter 1); glycolysis now ceases and all ATP-dependent functions come to a halt for lack of ATP.

The low and practically constant rate of post-mortem glycolysis apparent during the initial period after the blood supply has ceased, reflects the fact that the free Ca^{2+} concentration in the cytosol continues to be held well below the contractile threshold by ATP-dependent Ca^{2+} pumping, and this normally continues to be the case until the muscle fibre enters rigor mortis. In certain cases, however, the rate of glycolysis rises to higher levels. Thus, in the muscle of stress prone pigs, the rate of post-mortem metabolism is considerably enhanced, frequently more than ten-fold. This is apparently a result of somewhat elevated concentrations of Ca^{2+} free in the cytosol.[7] This condition, which, on the carcass, normally results in a low muscle pH whilst high temperatures still prevail, produces high drip losses and undesirably pale meat.[2]

Similarly, when the carcass is cooled prematurely to temperatures below 10°C, a shortening of some muscles occurs which is apparently the outcome of partial contractile activation caused by the prevailing low temperature.[2] This cold-shortening condition produces an accelerated rate of post-mortem metabolism[8] and is undesirable in commercial practice, particularly in beef and sheep carcasses, because it induces greater toughness in the meat which is produced.

Conversely, in certain circumstances, commercial practice may actually exploit the greatly accelerated rate of post-mortem metabolism which results from a bout of contractile activity. Electrical stimulation[9] is now widely used commercially to elicit massive contractile activity in the carcass shortly after death. The fuel reserves within the muscle are thereby rapidly depleted, the progress of events post mortem is greatly speeded and the muscle passes into rigor much more rapidly than normal. In this way one can eliminate the undesirable effects of the cold shortening which is normally elicited by premature cooling since the musculature is already ATP-depleted and unable to contract when it is subjected to the low temperature.

THE RECENT EXPERIMENTAL CLARIFICATION OF SOME PREVIOUSLY OBSCURE ISSUES

I have outlined above the major features of post-mortem metabolism. There are, however, some substantial gaps remaining in our understanding of the underlying phenomena; I now wish to devote my attention to certain of these gaps and to consider how far they have been filled by my recent experimental work. In order to introduce these considerations, I will pose the following three questions and deal with them in sequence.

1. Why is the onset of rigor stiffness in whole muscles rather protracted?

2. Is the phenomenon of cold shortening caused by an increase

in cytoplasmic free Ca^{2+} concentration?

3. Is the onset of rigor mortis associated with a Ca^{2+}-dependent contraction?

THE PROTRACTED ONSET OF RIGOR

As outlined in Fig.3, the onset of rigor stiffness in a mammalian muscle at 37°C may take more than an hour to reach its maximum value and the ATP content declines over the same period from 2-4 μmoles per g to practically zero. If the ATP were evenly distributed amongst the various muscle fibres this would imply that, as the ATP concentration declined from about millimolar towards zero, so the stiffness of each of the constitutent muscle fibres increased progressively. This is, however, at variance with experimental data derived from membrane-free single muscle fibres where it is possible experimentally to control the composition of the environment bathing the contractile filament lattice. Rigor does not begin to set in until the ATP concentration bathing the myofilaments is made very low (less than 10^{-4} M).[10],[11] If, however, single muscle fibres are permitted to go into rigor by metabolic depletion in rather the same way as does a whole muscle, then the onset of the rigor state is very much more rapid than in the case of whole muscle (Fig.4a). The behaviour of a small bundle of muscle fibres is intermediate between that of the single fibre and the whole muscle (Fig.4b). How may such differences be explained? It seems unlikely that the isolated single fibres are significantly damaged or otherwise unrepresentative of individual muscle fibres within a whole muscle. Moreover, the time which elapses before an isolated single fibre goes into rigor is quite variable from fibre to fibre, suggesting strongly that individual muscle fibres within a whole muscle enter rigor at different times. It seems likely that this is what causes the rather protracted time-scale of rigor onset in a whole muscle and it probably reflects the fact that individual fibres differ considerably in glycogen content.[12] In a single fibre, by contrast, the rapid onset of the rigor state may well reflect the rapid decline of the intracellular ATP concentration to zero when once it has fallen below a value ($\simeq 10^{-4}$ M) at which rigor stiffness begins to set in.

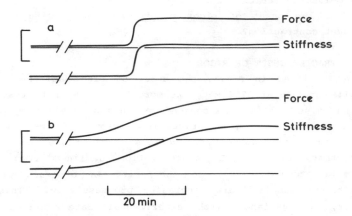

Figure 4 Time course of force and stiffness in a single beef sternomandibularis muscle fibre (a), and in a fibre bundle (b) 0.5 mm in diameter, derived from the same muscle. The recordings are interrupted for 4 hr at the point indicated and the light horizontal line below each of the experimental records denotes the values of zero force and zero stiffness. Scale bar on the ordinate corresponds to 200 mN mm^{-2} (force) and 40 N.mm^{-2} (stiffness) in (a), and 150 mN mm^{-2} (force) and 30 N.mm^{-2} (stiffness) in (b). The preparations were suspended in a physiological salt solution at 35°C with cyanide (1 mM) present to inhibit respiration and to ensure effectively anaerobic conditions. They were supported between a force transducer and an electromechanical displacement transducer which imposed upon the preparations small amplitude length changes at 77 Hz and thereby permitted continuous measurements to be made of muscle stiffness as well as of the force generated.

THE NATURE OF CONTRACTILE ACTIVATION IN COLD SHORTENING

The increasing evidence accumulated in recent years, for the central role played by free Ca^{2+} in the control of intracellular events, has stimulated much interest in experimental techniques for the measurement of the free Ca^{2+} concentration within living cells. Various techniques are now available for making this type of measurement directly[13] and the solution I have adopted is an optical one which exploits the colour changes caused when divalent cations bind to a particular dye, Arsenazo III, which can be injected into individual muscle fibres through a capillary microelectrode. The dye binds Ca^{2+} rather tightly (dissociation constant about 30 μM under intracellular conditions at pH 7) and Mg^{2+} rather weakly

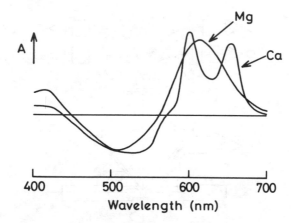

Figure 5 Difference spectra for the dye Arsenazo III generated by the addition of Ca^{2+} and Mg^{2+}.

(dissociation constant about 3 mM) to produce slightly different colour changes (Fig.5). By comparing measurements of dye absorbance at a wavelength where a large divalent cation-dependent change occurs, with measurements at another wavelength where the cation-dependent change is rather small, it is possible to distinguish between changes in dye absorbance and other potentially interfering changes in light transmission which, however, show only a weak dependence upon wavelength. This procedure is particularly necessary in muscle fibres where large changes in light scattering may occur during those changes in metabolic state which are associated with changes in free Ca^{2+}.

The microscope-based system actually employed is depicted in Fig.6. It permits simultaneous mechanical and absorbance measurements to be made and allows the single muscle fibre under observation to be heated and cooled very rapidly.

This technique has been used to study free Ca^{2+} changes in single mammalian muscle fibres during a cold contraction, as indicated in Fig.7. In this experiment, a beef muscle fibre, previously isolated, positioned in the apparatus and micro-injected with an approximately millimolar concentration of Arsenazo III, was caused to contract by cooling the bath from 12 to 3°C. The tension exerted by the fibre increased rather slowly and then rapidly declined once more when the higher temperature

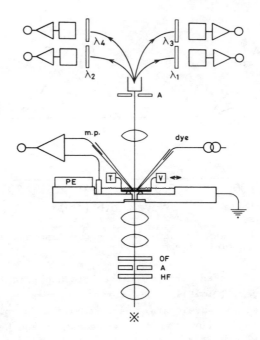

Figure 6 Diagram of the optical apparatus used to monitor absorbance changes at various wavelengths in single Arsenazo III-injected mammalian muscle fibres. Dye was injected into the single muscle fibre by electrophoresis via an Arsenazo III-filled microelectrode. The muscle fibre membrane potential was monitored with another intracellular microelectrode. Light, transmitted through a heat filter (HF), an aperture (A) and an orange filter (OF), was focussed to a spot 40 μm in diameter centered on the muscle fibre which was mounted between a force transducer (T) and an electro-mechanical vibrator (V) on the stage of a compound microscope. A long working distance microscope objective then collected the light which was focussed via an aperture (A) onto a bundle of glass fibre optic light guides which were arranged to direct the light simultaneously into one of up to four photomultiplier tubes, each protected by an interference filter transmitting maximally at a particular wavelength (λ_1 - λ_4). Electronic circuits permitted one of the amplified photomultiplier signals to be subtracted from each of the others. In this way, changes in light scattering could be eliminated from the absorbance record. The temperature of the preparation was set and could be altered with the aid of a servo-controlled peltier-effect refrigerator device (PE), muscle fibre tension was measured with the transducer (T) and minute alternations in fibre length from which the fibre stiffness could be inferred, could be applied via the electro-mechanical vibrator (V).

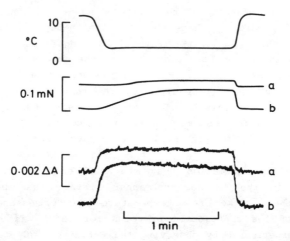

Figure 7 Time course of the mechanical and optical events in a dye-injected beef sternomandibularis fibre which was cooled from 12 to 3°C for a short period; upper record - time course of temperature change; middle record - tension before (a) and after (b) the addition of 0.5 mM caffeine to the bathing saline; lower record - absorbance change at 656 - 571 nm before (a) and after (b) the addition of 0.5 mM caffeine. The recording time constant was 0.2 sec. The intracellular concentration of Arsenazo III was 0.7 mM and the fibre diameter was 83 μm. Effectively anaerobic conditions were maintained by the presence of cyanide (1 mM) in the saline bathing the fibre. This treatment increases the amplitude of the cold contraction (14).

was re-established. The simultaneous optical record of the absorbance change at 656-571 nm showed respectively a rapid increase and a subsequent rapid decrease when the temperature was lowered and then raised once more. An upward deflection in the optical record when this wavelength pair is employed is characteristic of an increase in free Ca^{2+}. It therefore appears that when the fibre is cooled in this manner, cytoplasmic free Ca^{2+} does rise (the increment in this case in considerably less than 1 μm) and that this is associated with (rather weak) contractile activation. When a low concentration of caffeine is present, however, the strength of the contraction produced by the same drop in temperature is considerably enhanced and its onset is speeded, whilst the optical record indicates that the associated increase in free Ca^{2+} is also augmented. It thus appears that both the magnitude of the tension and the speed with

which it is generated are Ca^{2+}-dependent. Other similar
experiments indicate that the wavelength-dependence of the
absorbance change is Ca^{2+}- rather than Mg^{2+}-specific and it
is reasonable to conclude that cooling the muscle fibre in this
way causes (weak) contractile activation and force generation by
the release of Ca^{2+} (probably from the sarcoplasmic reticulum)
into the cytosol. This then explains the shortening observed in
unrestrained muscle <u>post mortem</u> if it is cooled to temperatures
below 10°C while it is still ATP-replete and competent to
contract.

A similar conclusion that the cold contraction is Ca^{2+}-
activated in the 'orthodox' manner (rather than being, for
example, the outcome of a temperature-dependent change in the
affinity of the contractile proteins for Ca^{2+}) was reached in
previous experiments by observing the efflux of ^{45}Ca from muscle
previously loaded with this isotope.[14] This latter technique,
however, allows only qualitative conclusions to be drawn
concerning the amplitude of the free Ca^{2+} change, and the time
resolution is also far inferior to that available in the optical
measurements made with Arsenazo III.

THE NATURE OF THE 'RIGOR CONTRACTION'

A muscle entering rigor develops considerable force and may
perform substantial amounts of mechanical work[2,15,16] if it
is permitted to shorten. It is likely that this ability of the
muscle fibre to shorten by several percent of its length and to
exert force during rigor onset is the result of the action of
transient force-generating cross-bridges which are established
between the thick and thin myofilaments and which operate via
cycles of attachment, pulling and detachment in the manner
described by Dr Davey (Chapter 1). The displacement produced by
a single working stroke of a cross-bridge is equivalent to less
than 1% of the muscle length. These transient force-generating
elements require the presence of ATP and they are finally
converted into the persistent interactions between the thick and
thin filaments which characterise ATP-free rigor muscle.

The muscle fibre is held relaxed by the plasticising effect
of ATP when the free Ca^{2+} ion concentration is kept
sufficiently low; activation and force generation occurs when the
free Ca^{2+} concentration rises to $\sim 10^{-6}$ M and rigor sets in if

the ATP falls below $\sim 10^{-4}$ M. The question thus arises: how is contractile activation produced in a muscle fibre which is entering rigor as a result of metabolic depletion in the normal way? Although at low ATP concentrations the Ca^{2+} sensitivity of the contractile apparatus rises somewhat,[10,11,17] it seems likely that the cytoplasmic free Ca^{2+} rises at the same time, since the normal low free Ca^{2+} concentration is only maintained by ATP-dependent Ca^{2+} pumps, the operation of which would be expected to be impeded as ATP declines. Some evidence in favour of this elevated Ca^{2+} hypothesis idea was obtained by Nauss and Davies using ^{45}Ca-loaded muscle.[18] More direct evidence might be forthcoming using the various direct Ca^{2+}-measuring techniques which are now available. Measurements with Arsenazo III, however, although they do show a very large increase in free Ca^{2+} after the onset of rigor, are unable to provide the necessary information because, at the time of rigor onset, a substantial increase in free Mg^{2+} occurs which effectively obscures the much smaller Ca^{2+} specific change.[19] Another (albeit indirect) technique does, however, provide assistance by showing that the mechanical characteristics of the rigor state generated naturally by metabolic depletion are similar to those of that rigor state which is produced 'artificially' by withdrawing ATP in the midst of a Ca^{2+}-activated contraction and are quite distinct from the rigor state produced by removing ATP from a relaxed muscle fibre.

The muscle fibre in rigor has mechanical properties which are very different from those of an ideal spring; in particular, the stiffness rises progressively as the force borne by the fibre increases; the fibre does not obey Hooke's Law.[20] This is demonstrated in Fig.8 which records the relationship between stiffness and tension for a particular single beef muscle fibre which was placed experimentally in three different rigor states. Initially, the fibre was allowed to go into rigor naturally by metabolic depletion and then, after chemically removing the muscle cell membranes with detergent, the two different 'artificial' rigor states described above were induced by removing ATP either in the presence or in the absence of sufficient Ca^{2+} to activate the contractile apparatus. It is clear that the mechanical characteristics of the 'natural' rigor state are very similar indeed to those of the 'artificial' state

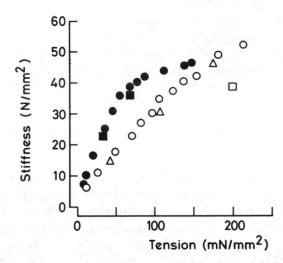

Figure 8 The dependence of stiffness upon tension exhibited by a single beef sternomandibularis muscle fibre in various rigor states. Open triangles denote the rigor state in the fibre when first it entered rigor naturally by metabolic depletion. Open circles denote the state subsequently produced in the same fibre after removing the cell membranes chemically with a detergent, causing contractile activation with Ca^{2+} and ATP and then producing rigor by ATP-withdrawal. Closed symbols denote the considerably stiffer state produced when the membrane-free fibre was relaxed with ATP in the virtual absence of Ca^{2+} and then rigor was induced by ATP withdrawal. The open square indicates the force and stiffness exhibited by the same fibre when maximally contracted in the presence of ATP and 10^{-5} M Ca^{2+}. The temperature was 35°C. Changes in the force borne by the rigor fibre were made by changing the length of the preparation.

which is produced by removing ATP from the maximally contracted muscle fibre. This observation clearly supports the idea that in the normal course of events, as the cellular ATP concentration declines to the point where rigor begins to set in, so sufficient Ca^{2+} is released into the cytoplasm to activate a contraction. The higher rate of energy expenditure then prevailing would be expected rapidly to scavenge away the last traces of ATP and produce rigor.

CONCLUSIONS

   The   theme   which   clearly   pervades   this   description   of metabolic events <u>post mortem</u> is the all important control function performed by the Ca^{2+} ion. The results of the experiments which I have described serve to indicate that in two different areas of post-mortem metabolism - during the cold contraction of an ATP-replete muscle and during the onset of <u>rigor mortis</u> itself - Ca^{2+}-regulated phenomena determine both the tempo and the mode of these events.

REFERENCES

1. D.G. Gadian, G.K. Radda, T.R. Broom, E.M. Chance, M.J. Dawson and D.R. Wilkie, <u>Biochem.J.</u>, 1981, <u>194</u>, 215.
2. J.R. Bendall, "The Structure and Function of Muscle", Academic Press, New York, 1973, Vol.2, p.243.
3. C.O. Brostrom, F.L. Hunkeler and E.G. Krebs, <u>J.Biol.Chem.</u>, 1971, <u>246</u>, 1961.
4. P. Cohen, "The Control of Enzyme Activity", Chapman and Hall, London, 1976, p.32.
5. R.M. Denton and J.G. McCormack, <u>FEBS Lett.</u>, 1980, <u>119</u>, 1.
6. R.E. Jeacocke, <u>FEBS Lett.</u>, 1982, <u>147</u>, 225.
7. J.N. Lucke, G.M. Hall and D. Lister, <u>Ann.N.Y.Acad.Sci.</u>, 1979, <u>317</u>, 326.
8. R.E. Jeacocke, <u>J.Sci.Food Agric.</u>, 1977, <u>28</u>, 551.
9. J.R. Bendall, C.C. Ketteridge and A.R. George, <u>J.Sci.Food Agric.</u>, 1976, <u>27</u>, 1123.
10. J.P. Reuben, P.W. Brandt, M. Berman and H. Grundfest, <u>J.Gen.Physiol.</u>, 1971, <u>57</u>, 385.
11. P.W. Brandt, J.P. Reuben and H. Grundfest, <u>J.Gen.Physiol.</u>, 1972, <u>59</u>, 305.
12. C.S. Hintz, M.M.-Y. Chi, R.D. Fell, J.L. Ivy, K.K. Kaiser, C.V. Lowry and O.H. Lowry, <u>Am.J.Physiol.</u>, 1982, <u>242</u>, C, 218.
13. J.R. Blinks, W.G. Wier, P. Hess and F.G. Prendergast, <u>Prog. Biophys.Molec.Biol.</u>, 1982, <u>140</u>, 1.
14. R.E. Jeacocke, <u>Biochim.Biophys.Acta</u>, 1982, <u>682</u>, 238.
15. B.B. Marsh, <u>J.Sci.Food Agric.</u>, 1954, <u>5</u>, 70.
16. J.R. Bendall, <u>J.Physiol.,Lond.</u>, 1951, <u>114</u>, 71.
17. R.E. Godt, <u>J.Gen.Physiol.</u>, 1974, <u>63</u>, 722.
18. K.M. Nauss and R.E. Davies, <u>J.Biol.Chem.</u>, 1966, <u>241</u>, 2918.
19. R.E. Jeacocke (In preparation).
20. R.E. Jeacocke, <u>Meat Sci.</u>, 1983 (In press).

4
Phosphorus NMR Studies of Muscle Metabolism

By David E. Gadian
PHYSICS IN RELATION TO SURGERY, ROYAL COLLEGE OF SURGEONS OF ENGLAND,
LINCOLN'S INN FIELDS, LONDON WC2A 3PN, U.K.

INTRODUCTION

Phosphorus nuclear magnetic resonance (^{31}P NMR) was first used to study cellular and tissue metabolism about ten years ago,[1,2] and since that time there has been increasing interest in this non-invasive method of studying metabolism in vivo (for reviews see [3-6]). In this article, some examples will be given of the ways in which ^{31}P NMR can be used to study muscle metabolism, and it is hoped that these provide an indication of some potential uses of the technique in studying meat chemistry. In addition, some potential applications using ^{1}H and ^{13}C NMR are discussed.

THE PRINCIPLES OF NMR

A detailed description of the principles of NMR is far beyond the scope of this article. However, it is hoped that the following brief discussion will be of value to any newcomers to NMR (for reviews see Gadian[7],[8]).

NMR is a branch of spectroscopy, and in common with other spectroscopic techniques, it detects the interaction of radiation with matter. The technique relies on the fact that certain atomic nuclei such as hydrogen (the proton, ^{1}H), carbon (^{13}C), and phosphorus (^{31}P), have intrinsic magnetic properties. When a sample containing such nuclei is placed in a magnetic field, the nuclei tend to align along the direction of the field. This magnetic interaction can be detected by applying radiofrequency radiation to the sample and observing frequencies at which radiation is absorbed and subsequently re-emitted. The frequencies of the various emitted signals are proportional to the field that is applied, but they also depend to a much lesser extent on the chemical properties of the molecules containing the nuclei. So, for example, the ^{1}H NMR spectrum of acetic acid, shown in Fig.1, contains two signals, or resonances, one from the three chemically equivalent CH_3 protons and another from the COOH proton. Each proton contributes an equal amount of signal, and so the relative intensities of the resonances, given by the areas under the peaks, are 3:1. The frequencies of the signals are expressed in terms of a dimensionless parameter known as the chemical shift, and are generally expressed in terms of parts per million (ppm) relative to the shift of a standard. The existence of the chemical shift enables NMR to provide information about the structure of molecules in solution, and the technique is routinely used by chemists and biochemists for studies of this type.

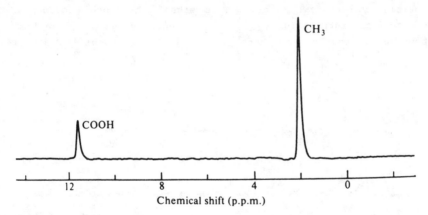

Figure 1 ^{1}H NMR spectrum of acetic acid. The frequencies of the signals are expressed in parts per million (ppm) relative to the signal from the reference compound tetramethylsilane (TMS).[8]

METABOLIC STUDIES

In 1973 and 1974, it was shown that ^{31}P NMR signals could be observed, not only from solutions, but also from metabolites in intact cells and in tissues isolated from animals. Different phosphorus-containing metabolites produce signals at different frequencies and simply by monitoring how these signal intensities vary with time, it is possible to observe the inter-conversions of the various molecules, i.e. to follow their metabolism. For example, Fig.2 shows spectra of a rat vastus lateralis muscle recorded at various times after excision of the muscle. The

spectra contain signals from the three (α, β and γ) phosphates of
ATP, phosphocreatine, inorganic phosphate, and sugar phosphates.
The simplicity of the spectra reflect the fact that well-resolved
signals are observed only from mobile phosphorus-containing
compounds that are present at concentrations of about 0.5 mM and
above.   In this experiment the muscle was in a closed NMR tube
throughout the period, and owing to lack of oxygen, the high-
energy compound phosphocreatine was gradually depleted, with a
corresponding formation of inorganic phosphate.   When the pool of
phosphocreatine is almost exhausted, the ATP level starts to
decline.

An important feature of ^{31}P NMR is that the frequency of the
inorganic phosphate signal is sensitive to pH variations in the
normal physiological range.   This signal therefore provides a
monitor of intracellular pH.   The sensitivity to pH arises
because inorganic phosphate has a pK_a of about 6.75 and

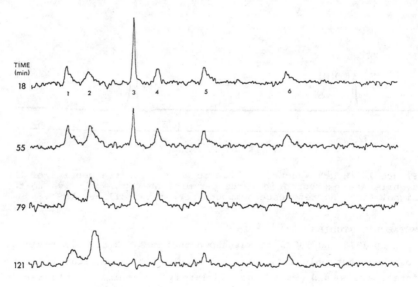

Figure 2 ^{31}P NMR spectra of a vastus lateralis muscle from the
hind leg of a rat.   The temperature was 20°C.   The signals are
assigned as follows: 1, sugar phosphates and AMP/IMP; 2,
inorganic phosphate; 3, phosphocreatine; 4, 5, and 6, the γ, α
and β phosphates of ATP.   The spectra were recorded at 129 MHz,
and the spectral width is 5 kHz. (From Hoult et al.[2] Reprinted
by permission from <u>Nature</u>.)

therefore its state of ionisation changes in the physiological pH
range. Thus in the spectra of Fig.2, the inorganic phosphate
signal shifts gradually to the right, indicating a decline in
average pH from about 7.1 to 6.5, presumably due to lactic acid
accumulation. The width and shape of this signal probably
reflects the distribution of pH values throughout the tissue.

This type of study can be extended by maintaining isolated
muscles under good physiological condition in the spectrometer.
In fact NMR provides an ideal method of relating metabolic state
to physiological function as exemplified by some ^{31}P NMR studies
of muscular fatigue, a subject about which surprisingly little is
known.

Fig.3 shows the results of studies in which isolated frog

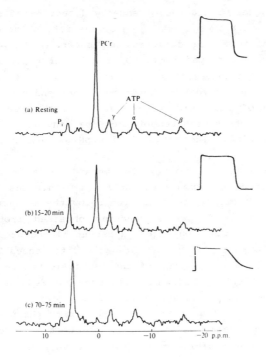

Figure 3 ^{31}P NMR spectra obtained from anaerobic frog
gastrocnemius muscles at 4°C during a fatiguing series of 5 sec
contractions repeated every 5 min. Adjacent to the spectra are
mechanical records showing the time course of isometric force
development; the force progressively declines and relaxation
becomes slower. (Adapted from Dawson <u>et al</u>.[9])

gastrocnemius muscles were stimulated repetitively for 5 seconds
every 5 minutes.   The force measurements shown at the right of
each spectrum reveal the typical signs of fatigue; the force
gradually declines in magnitude and the rate of mechanical
relaxation becomes slower.   The spectra indicate the expected
decline in phosphocreatine, increase in inorganic phosphate, and
decline in pH due to the formation of lactic acid.   The
correlations that were observed between the metabolic state and
the tension response provided new information about the metabolic
basis of muscular fatigue.[9,10]

In 1979, it was shown that excellent ^{31}P NMR spectra could
be observed from selected regions of live animals such as the
rat.[11]   For example, an anaesthetised rat can be positioned
within the magnet with a radiofrequency detecting coil placed in
the required position against a leg of the rat.   Remarkably clear
^{31}P NMR signals can be detected, as shown in Fig.4, from an
approximately disc-shaped region of the leg immediately in front
of the coil.   One interesting observation from Fig.4a is the very
small size of the inorganic phosphate signal indicating that the
concentration of inorganic phosphate within the muscle is very
much lower than that measured by the more conventional technique
of freeze clamping.   The intracellular pH as measured by NMR is
about 7.1.

The spatial selectivity of the method makes it possible to
assess the effects of localised ischaemia within the leg muscle.
Fig.4a shows a spectrum obtained from below the knee under normal
conditions.   Fig.4b shows a spectrum obtained from the same
location after a tight tourniquet had been placed around the leg
just above the knee.   There are dramatic differences between the
two spectra, for the spectrum of the ischaemic muscle shows that
there is very little phosphocreatine, somewhat reduced amounts of
ATP, and a very large concentration of inorganic phosphate.   In
addition the frequency of the inorganic phosphate signal
indicates a decline in pH to 6.7.   A spectrum obtained from leg
muscle above the tourniquet (Fig.4c) is very similar to that
shown in Fig.4a, confirming that the muscle above the knee has
not suffered from ischaemia.

With the more recent development of large magnets, studies
of human limbs are now feasible.[12-16]   Fig.5a shows a ^{31}P NMR
spectrum from tissue within the flexor compartment of the forearm

of a healthy subject. Again, the level of inorganic phosphate is much lower than values obtained by the technique of needle biopsy, presumably, at least in part, because this latter method involves an unavoidable breakdown of high-energy phosphates prior to analysis. The intracellular pH, as indicated in the Figure, is measured to be 7.04. On ischaemic exercise, large changes are observed, as shown in Fig.5b and c, and following the restoration of arterial flow, the metabolic state recovers to normal in a few minutes. Similar changes can be observed during normal exercise – indeed the pH can sometimes fall to as low as 6.0.[13,16]

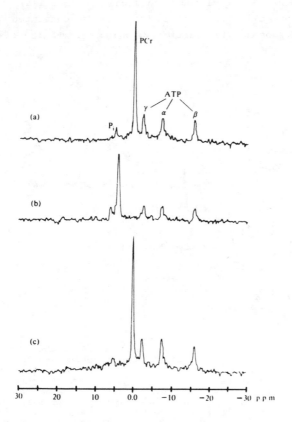

Figure 4 ^{31}P NMR spectra obtained at 73.8 MHz from the leg of an anaesthetised rat. Each spectrum represents the accumulation of 128 scans repeated at 12 sec intervals. (From Ackerman et al.[11] Reprinted by permission from Nature.)

Studies of this type provide a great deal of information about the metabolism of healthy muscle, and in addition they provide the baseline for studies of muscle disease.

Fig.6 shows spectra recorded in a similar manner from a patient suspected as having McArdle's syndrome. This is an inborn error of metabolism caused by a lack of glycogen phosphorylase activity in skeletal muscle, and is diagnosed by the demonstration that ischaemic exercise fails to generate lactic acid. The main feature that distinguishes the spectra of Fig.6 from those of Fig.5 is that there is no decrease in intracellular pH associated with ischaemic exercise; in fact the pH increases slightly. This is entirely consistent with the absence of phosphorylase activity, and confirmed the diagnosis of

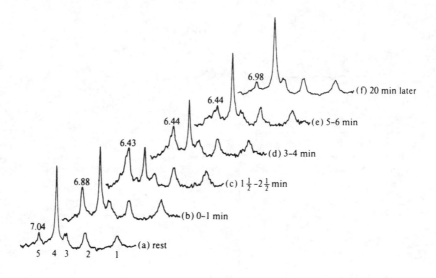

Figure 5 ^{31}P NMR spectra obtained at 32.5 MHz from a human forearm. The first spectrum (a) was recorded at rest, prior to exercise; subsequent spectra (b)-(f) were recorded during the periods shown, where 0 min corresponds to the time at which ischaemic exercise was started. Exercise was maintained during the period 0-1½ min, but arterial occlusion was maintained up to 3 min. Arterial flow was restored after this period. The signals are assigned as follows: 1, 2 and 3, the β-, α- and γ-phosphates of ATP; 4, phosphocreatine; 5, inorganic phosphate. Measured pH values are given above each inorganic phosphate signal. (From Ross et al.[12] Reprinted by permission of New Eng.J.Med.)

McArdle's disease; the absence of phosphorylase was then firmly established by direct enzymatic assay of a muscle biopsy.

SOME ASPECTS OF NMR INSTRUMENTATION AND TECHNOLOGY – NMR IMAGING

The single most important feature of an NMR spectrometer is probably the magnet, which has to be designed to produce a highly uniform magnetic field. For metabolic studies, it is often necessary to compromise between the strength of the magnetic field, the size of the bore into which the sample is placed and other considerations, including finances. Thus the most conventional NMR spectrometers have magnets that generate a high field (about 4–12 Tesla) but have a limited bore size (5–10 cm diameter). Spectrometers of this type are most commonly used for solution studies, but they can also be used for studies of cellular suspensions, isolated tissues and, as described above, small animals. Larger bore (20–60 cm diameter) magnets of lower

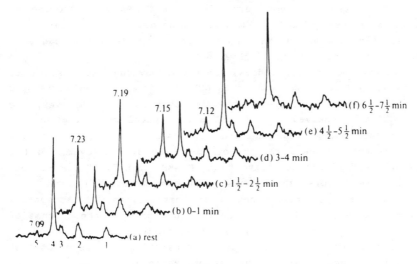

Figure 6 ^{31}P NMR spectra obtained at 32.5 MHz from a patient with McArdle's syndrome. The first spectrum (a) was recorded at rest, prior to exercise; subsequent spectra (b)-(f) were recorded during the periods shown, where 0 min corresponds to the time at which ischaemic exercise was started. Exercise was maintained during the period 0-¾ min, but arterial occlusion was maintained up to 3 min. Arterial flow was restored after this period. The signal assignments are as for Fig.5. Measured pH values are given above each inorganic phosphate signal. (From Ross et al.[12] Reprinted by permission of New Eng.J.Med.)

field strength (1.5-1.9 Tesla) have more recently become available and these are suitable for studying the metabolism of relatively large samples, e.g. human beings! In the future, it should also be possible to perform NMR imaging studies on such instruments; for various reasons, most imaging studies are at present performed at much lower field strengths. Although imaging is not the subject of this article, it is appropriate to discuss, albeit very briefly, this rapidly developing NMR technique, for it could prove to be of considerable relevance to meat science.

In NMR imaging, ^{1}H signals are observed from the protons of water within samples, and by the use of magnetic field gradients, it is possible to map out the 2- or 3-dimensional distribution of the protons from which the signal is observed.[17,18] In this way, images can be generated that are analogous in some respects to those obtained by X-ray CT scanning. The technique has produced some extremely impressive human images, particularly of the head;[17-19] and it seems very likely that NMR will provide an invaluable new method of medical imaging. One particularly interesting feature of NMR imaging is that information can also be obtained about rates of flow.

For biochemical studies, it would obviously be useful to combine NMR imaging with ^{31}P NMR in order to evaluate the metabolic state of different regions of a sample. However, the metabolites observed by ^{31}P NMR have typical concentrations of about 1-20 mM, whereas the protons observed in imaging studies are present at concentrations of up to 100 M. Moreover, ^{31}P NMR is inherently less sensitive than ^{1}H NMR, and for these reasons there are severe sensitivity problems associated with spin imaging of phosphorus-containing metabolites. Acceptable signal-to-noise ratios could be obtained only at the expense of dramatic reduction in spatial resolution; thus one would be limited to relatively crude images, consisting perhaps of at most 20 volume elements within the region examined.

It is mainly for this reason that spatial resolution for most metabolic studies has, to date, been achieved not by imaging but rather by investigating the metabolic state of a single selected, localised region within a sample. Two methods of localisation have been successfully used for ^{31}P NMR studies; one of these involves 'surface' radiofrequency coils, which are used for the animal and human studies of the type described above. The second method, termed topical magnetic resonance[20] involves a special form of magnetic field profiling, and has been used to localise on internal regions, in some cases in conjunction with surface coils. In this way, signals have been observed from the liver[21] and from the kidney[22] of intact living rats.

APPLICATIONS IN MEAT CHEMISTRY

The above examples illustrate the metabolic changes that can be observed during periods of ischaemia, exercise, and recovery, and it is hoped that they suggest some of the potential applications of NMR in studies of meat chemistry. For example, it may be possible to use ^{31}P NMR to investigate the biochemical processes resulting from electrical stimulation of carcasses or isolated muscles soon after slaughter of animals. Unfortunately, it is not easy to evaluate here the future role of NMR in such studies, partly because, to the author's knowledge, so far there have been very few publications in this general area of research. Some work has been done on the metabolism of polyphosphates which are used as food additives by the frozen meat and fish industries to reduce water losses during processing, thawing, and cooking. In the case of frozen poultry, an aqueous solution of polyphosphate salts is commonly injected into the carcass breast muscle before it is chilled and frozen. These salts produce ^{31}P NMR signals in regions fairly close to the β and γ phosphate signals of ATP, and can therefore be clearly distinguished from orthophosphates. Among the results obtained by these studies[23-25] it was shown that substantial hydrolysis of the added polyphosphates took place in frozen chicken during long-term storage (up to 43 months) at $-20°C$. It was also shown that when an aqueous solution of sodium tripolyphosphate was injected into excised chicken breast muscle, sequential hydrolysis of the tripolyphosphate could be observed over a period of several hours. Such studies can clarify the biochemical effects of polyphosphate treatment.

THE USE OF OTHER NUCLEI

A considerable number of relevant studies has been performed using nuclei other than [31]P. For example, ^{1}H, [23]Na and, to a very limited extent, [39]K NMR have been used to investigate the properties of water and monovalent cations in isolated muscle.[26] However, it is probably fair to say that in view of the effort that has been put into these studies, the amount of conclusive information that has emerged is a little disappointing.

In addition, preliminary high-resolution ^{13}C NMR[27] and ^{1}H NMR[28] studies have shown that it may be possible to use these nuclei to study certain aspects of the metabolism of isolated intact muscle. Indeed, ^{13}C and ^{1}H NMR could prove to be at least as valuable as ^{31}P NMR in studies of meat chemistry. For example, ^{13}C NMR has been used to determine 4-hydroxy-L-proline in meat protein.[29,30] This amino acid is found at appreciable levels only in collagen, and is therefore an accepted indicator for connective tissue. It is interesting to note that ^{13}C NMR signals can also be observed from triglycerides in the human forearm.[15,31,32] In addition, high resolution ^{1}H NMR spectra of the forearm show two main peaks, one from water and the other from fat,[15,16,32] and ^{1}H and ^{13}C NMR could therefore provide useful information about the composition and relative amounts of the tissue that is being examined. If it proves possible to combine such high resolution studies with ^{1}H spin-imaging, further information may be available about the location of muscle, fat and connective tissue within the region under investigation.

ACKNOWLEDGEMENTS

The ^{31}P NMR studies described in this article were performed in the Biochemistry Department of the University of Oxford with support from the Medical Research Council, Science and Engineering Research Council, and the British Heart Foundation. The author thanks the Rank Foundation for support at the Royal College of Surgeons of England.

REFERENCES

1. R.B. Moon and J.H. Richards, J.Biol.Chem., 1973, 248, 7276.
2. D.I. Hoult, S.J.W. Busby, D.G. Gadian, G.K. Radda, R.E. Richards and P.J. Seeley, Nature, 1974, 252, 285.
3. D.G. Gadian, Ann.Rev.Biophys.Bioeng., 1983, 12, 69.
4. D.G. Gadian and G.K. Radda, Ann.Rev.Biochem., 1981, 50, 69.
5. J.R. Griffiths, R.A. Iles and A.N. Stevens, "Progress in NMR Spectroscopy", Pergamon Press, Oxford, New York, 1982,

Vol.15, p.49.

6. R.G. Shulman, Scientific American, 1983, 248, 76.

7. D.G. Gadian, "Developments in Meat Science", Applied Science Publishers, London, 1980, Vol.1, p.89.

8. D.G. Gadian, "Nuclear Magnetic Resonance and its Applications to Living Systems", Oxford University Press, Oxford, 1982.

9. M.J. Dawson, D.G. Gadian and D.R. Wilkie, J.Physiol., 1980, 299, 465.

10. M.J. Dawson, D.G. Gadian and D.R. Wilkie, Nature, 1978, 274, 861.

11. J.J.H. Ackerman, T.H. Grove, G.G. Wong, D.G. Gadian and G.K. Radda, Nature, 1980, 283, 167.

12. B.D. Ross, G.K. Radda, D.G. Gadian, G. Rocker, M. Esiri and J. Falconer-Smith, New Engl.J.Med., 1981, 304, 1338.

13. G.K. Radda, L. Chan, P.J. Bore, D.G. Gadian, B.D. Ross, P. Styles and D. Taylor, "NMR Imaging", Bowman Gray School of Medicine of Wake University, Winston, Salem, NC, USA, 1982, p.159.

14. G.K. Radda, P.J. Bore, D.G. Gadian, B.D. Ross, P. Styles, D.J. Taylor and J. Morgan-Hughes, Nature, 1982, 295, 608.

15. R.H.T. Edwards, M.J. Dawson, D.R. Wilkie, R.E. Gordon and D. Shaw, Lancet, 1982, i, 725.

16. D.J. Taylor, P.J. Bore, P. Styles, D.G. Gadian and G.K. Radda, Mol.Biol.Med., in press.

17. I.L. Pykett, Scientific American, 1982, 246, 78.

18. R.L. Witcofski, N. Karstaedt and C.L. Partain, "NMR Imaging", Bowman Gray School of Medicine of Wake University, Winston Salem, NC, USA, 1982.

19. G.M. Bydder, R.E. Steiner, I.R. Young, A.S. Hall, D.J. Thomas, J. Marshall, C.A. Pallis and N.J. Legg, Am.J.Roentgenology, 1982, 139, 215.

20. R.E. Gordon, P.E. Hanley and D. Shaw, "Progress in NMR Spectroscopy", Pergamon Press, Oxford, New York, 1982, Vol.15, p.1.

21. R.E. Gordon, P.E. Hanley, D. Shaw, D.G. Gadian, G.K. Radda, P. Styles, P.J. Bore and L. Chan, Nature, 1980, 287, 736.

22. R.S. Balaban, D.G. Gadian and G.K. Radda, Kidney Int., 1981, 20, 575.

23. I.K. O'Neill and C.P. Richards, Chem.Ind., 1978, (Jan), 65.

24.   M. Douglass, M.P. McDonald, I.K. O'Neill, R.C. Osner and
      C.P. Richards, J.Food Technol., 1979, 14, 193.

25.   P.J. King and C.P. Richards, Bull.Magn.Reson., 1981, 2,
      383.

26.   M. Shporer and M.M. Civan, Curr.Top.Memb.Trans., 1977, 9, 1.

27.   D.D. Doyle, J.M. Chalovich and M. Barany, FEBS Lett., 1981,
      131, 147.

28.   K. Yoshizaki, Y. Seo and H. Nishikawa, Biochim.Biophys.Acta,
      1981, 678, 283.

29.   M.L. Jozefowicz, I.K. O'Neill and H.J. Prosser,
      Analyt.Chem., 1977, 49, 1140.

30.   I.K. O'Neill, M.L. Trimble and J.C. Casey, Meat Sci., 1978,
      3, 223.

31.   J.R. Alger, L.O. Sillerud, K.L. Behar, R.J. Gillies and
      R.G. Shulman, Science, 1981, 214, 660.

32.   R.E. Gordon, Phys.Bull., 1981, 32, 178.

5
Water-holding in Meat

By Gerald Offer, David Restall, and John Trinick
A.R.C. MEAT RESEARCH INSTITUTE, LANGFORD, BRISTOL BS18 7DY, U.K.

INTRODUCTION

Fresh meat at slaughter contains about 75% water. But the amount is not fixed and there can be very substantial changes depending on the treatment the meat receives post-slaughter.[1-4] Losses of water from the meat (which would lead to shrinkage) can occur from three causes. One is evaporation from the surface of the meat where the loss is typically of the order of 3% but in bad commercial practice can reach 12%. Another cause is drip from the cut ends of meat where the loss is again typically about 3% but is exacerbated in the PSE condition or after freezing and thawing when it can reach 15%. Finally, in cooking, losses occur that are even greater and can be as high as 40%.

On the other hand, sodium chloride is commonly used to promote water uptake and consequent swelling in the manufacture of meat products such as sausages, burgers or tumbled hams.[1-4] The amount of water uptake may be as low as 5% but can reach 40% in a massaged ham. Small amounts of polyphosphate are often also added.[1-7] Longer chain phosphates such as triphosphate and tetraphosphate are broken down in the meat and it is believed that pyrophosphate is the active substance.[8] The purpose of adding polyphosphate is firstly to reduce the amount of sodium chloride required and secondly to create conditions where meat pieces bind together. By massaging or tumbling meat in the presence of sodium chloride and polyphosphate a sticky exudate is formed on the surface of the meat pieces which acts as a cement, particularly after cooking.[9,10] It is thought that myosin is the agent responsible.

Gains or losses of water in meat are important for two reasons. They are important economically since meat is sold by weight. They are also important for consumer satisfaction since they affect the juiciness, texture and flavour of meat. Despite their importance, the literature on water-holding is not at all

straightforward.   In this chapter I want to clarify the problem
of water-holding by asking two questions:

(1)   **Where** in meat is water held, where is it taken up and from
where is water lost?

(2)   **How?**: that is, what is the mechanism of water loss or gain?

    A muscle fibre contains a thousand or more myofibrils packed
closely together.[11-12] Within each myofibril, the thick and thin
filaments are arranged in a hexagonal lattice, a thin filament
being at the centre of three thick filaments.   It is worth
emphasising that a very high fraction (80-87%) of the fibre
volume is occupied by myofibrils.[13]   Thus myofibrils are the
largest water compartment in muscle.   This being so, a highly
attractive hypothesis can be proposed: **all changes in water
holding are due to changes in the volume of myofibrils** (Fig.1).
Gains in water leading to swelling of meat would be due to
swelling of the myofibrils caused by expansion of the filament
lattice, while losses in water would be due to shrinking of the
myofibrils.

WATER UPTAKE

    Let us first consider whether this hypothesis is true for
water uptake.   This may be tested by examining myofibrils,
prepared by homogenising meat in a suitable buffer, in the wet
state in the phase-contrast light microscope.   This type of
approach was pioneered by Hanson and Huxley in their classical
work on muscle contraction.[14]   One can measure changes in the
volume of myofibrils when the buffer is changed by following

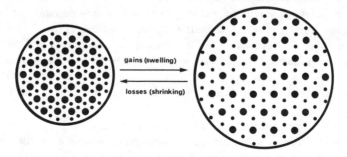

Figure 1 Hypothesis explaining changes in water holding.   On
the left hand side is shown schematically a transverse section of
a myofibril.   On the right hand side the same myofibril is shown
with an expanded filament lattice.

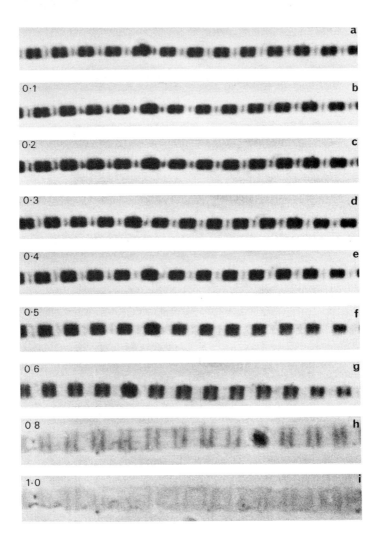

Plate 1. Swelling of a myofibril in sodium chloride.
The same rabbit psoas myofibril is shown (a) in the preparation medium, (0.1 M KCl 2 mM MgCl$_2$ 1 mM EGTA 0.5 mM
dithiothreitol 10 mM potassium phosphate pH 7.0) then
after successive 3 minute irrigations with 1 mM MgCl$_2$,
10 mM sodium acetate, pH 5.5 containing the following
NaCl concentrations (M): (b) 0.1 (c) 0.2 (d) 0.3
(e) 0.4 (f) 0.5 (g) 0.6 (h) 0.8 (i) 1. Magnification
4000 x.

Reproduced from <u>Meat Science</u> with permission

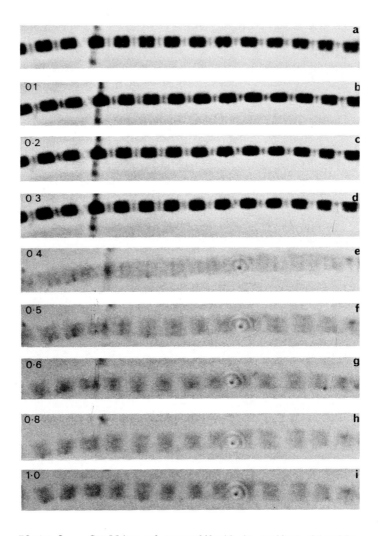

Plate 2.   Swelling of a myofibril in sodium chloride
plus pyrophosphate.   The experiment was performed in
exactly the same way as in Plate 2 but in (b) to (i)
10 mM sodium pyrophosphate was also present in the
irrigating solutions.

Reproduced from <u>Meat Science</u> with permission

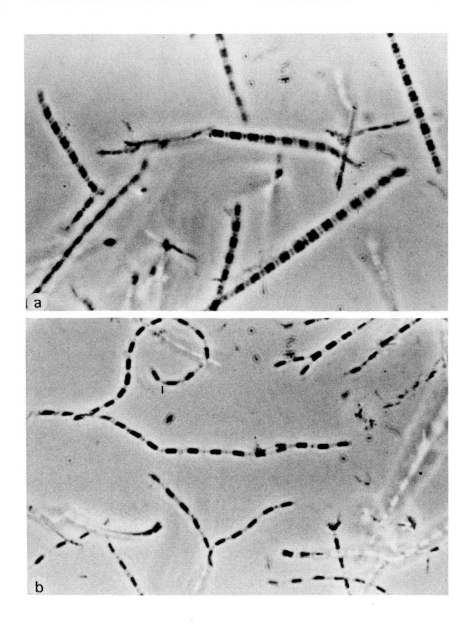

Plate 3.   The effect of heating on myofibril volume.   (a) rabbit
psoas myofibrils in the rigor state, (b) after heating at 61°C for
30 minutes in 0.1 M NaCl, 1 mM MgCl$_2$, 10 mM sodium acetate pH 5.5.

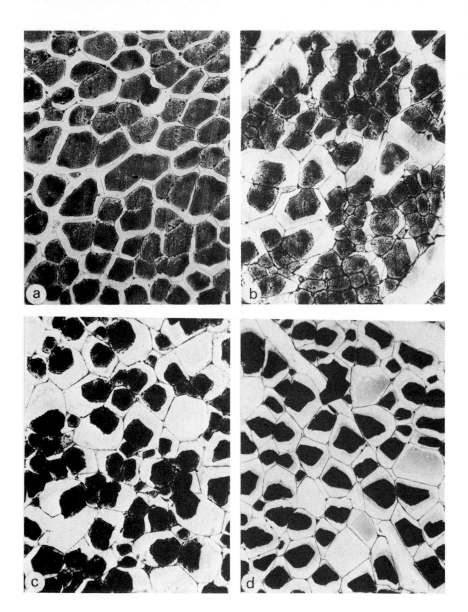

Plate 4.   Transverse sections of beef psoas muscle before and
after heating stained with Gomori's reticulin method.
(a) unheated muscle in the rigor state (b) after heating at 40°C
(c) 50°C (d) 60°C (e) 70°C (f) 80°C (g) 90°C. Magnification 250 x.

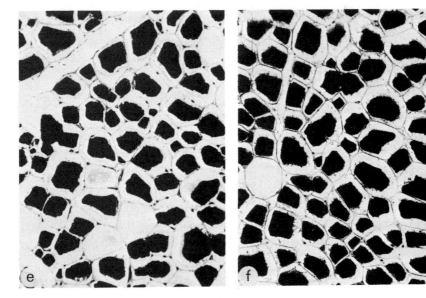

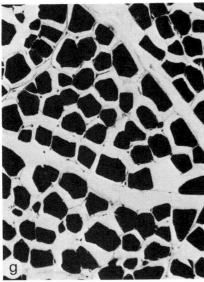

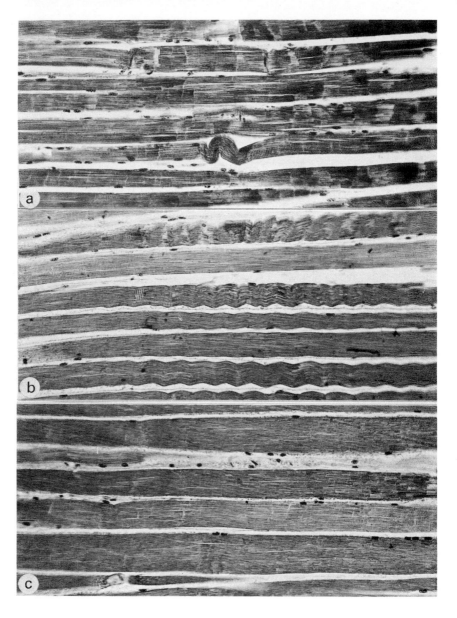

Plate 5.   Longitudinal sections of beef psoas muscle before and
after heating.   (a) Unheated muscle in the rigor state.
(b) After heating at 70°C (c) 90°C.   Magnification 330 x.

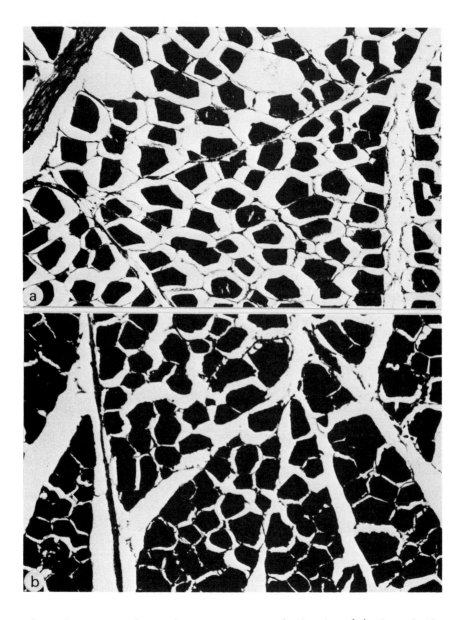

Plate 6.   Comparison of transverse sections taken (a) through the
middle and (b) near the cut ends of the block of beef
sternomandibularis muscle heated at 80°C for 30 minutes.

changes in the diameter of the myofibril and in its sarcomere length. One can also study changes in the band pattern of the myofibrils and thus detect whether proteins are extracted. A detailed account of our work on water uptake will be found in reference 15.

Plate 1 shows the appearance of a myofibril after successive 3 minute treatments with buffers containing an increasing concentration of sodium chloride. In the top frame is a myofibril in its preparation medium at pH 7.0. In the second frame the same myofibril is shown after treatment with a solution lowering the pH (0.1 M NaCl, 1 mM $MgCl_2$, 10 mM sodium acetate, pH 5.5); a slight shrinkage occurs due to this fall in pH. As the sodium chloride concentration is increased, little change occurs (except for the disappearance of the Z-line) until a concentration of 0.6 M is reached. At this point the centre of the A-band is extracted and the myofibrils swell dramatically. After treatment with buffers containing 0.8 and 1.0 M NaCl, more of the A-band is extracted and further swelling takes place. The sarcomere length is unchanged by the treatment. In the example shown the diameter has increased to 2.5 times the original value. If we assume the myofibril is cylindrical, the volume is proportional to the square of the diameter and thus this myofibril has swollen by the remarkable value of six times.

We find considerable variation from one preparation of myofibrils to another. We have always observed swelling in high concentrations of sodium chloride but the degree of swelling is variable, a two-fold expansion in volume being more typical. This is, of course, of particular interest to the meat industry which experiences a variation in the suitability of meat samples for processing. We also find variation in the appearance of the Z-line after treatment. Sometimes, the Z-line swells considerably, and, as in the example of Plate 1, it may disappear. In such cases the swelling of the A-band and I-band are similar and the myofibrils remain approximately cylindrical. Sometimes, however, the Z-line does not swell. In such cases the A-band swells much more than the I-band and the myofibril is pinched in at every Z-line. Finally, we observe variation in the extraction of the A-band; quite commonly with sodium chloride alone, we observe little extraction of the A-band.

Such variations should not distract us from the main

conclusions:

1. Myofibrils swell in high concentrations of sodium chloride.

2. The degree of swelling is more than enough to explain the water uptake in meat processing.

3. The concentration of sodium chloride required for swelling is similar to that found for comminuted meat.

Similar experiments may be performed but now including a constant level of 10 mM pyrophosphate, a concentration similar to that used in meat processing (Plate 2). As before, a slight shrinkage occurs after the myofibril is irrigated with the buffer at pH 5.5. No change occurs in solutions containing up to 0.3 M NaCl; indeed irrigation with the solution containing 0.3 M NaCl may be prolonged indefinitely without effect. But when the myofibril is irrigated with a medium containing 0.4 M NaCl, very dramatic changes occur. The A-bands are extracted beginning from their outer edges; after 3 minutes irrigation they are completely extracted leaving strings of I-bands. As the extraction begins, swelling occurs. The process is highly co-operative, the increase in sodium chloride concentration from 0.3 to 0.4 M in the presence of pyrophosphate having a profound effect.

Thus with pyrophosphate the concentration of sodium chloride for maximum swelling is halved; this nicely demonstrates the synergistic effect of pyrophosphate first emphasised by Bendall.[5] There appears to be no great effect of pyrophosphate on the maximum degree of swelling, although this needs to be tested more carefully with greater numbers of myofibrils. Most importantly, in the presence of pyrophosphate, the A-band is completely extracted; in meat processing such extraction would provide the myosin required for cementing meat pieces together.

These experiments vividly demonstrate that the most likely candidate for the site of water retention in meat processing is the myofibril.

We turn now to the second question posed: what is the mechanism of myofibrillar swelling? We would like at this point to acknowledge the debt we owe to Reiner Hamm for the major contributions he has made over three decades to our ideas on water holding.[1-3] However, although we have drawn heavily on these ideas, they are expressed in physico-chemical rather than structural terms and we believe that a structural explanation is needed for a complete understanding.

For myofibrils to change volume, the spacing of the filament lattice must alter (Fig.1). It is known from the pioneering work of Gerald Elliott and his colleagues using X-ray diffraction that the interfilament spacing can indeed change substantially when the medium bathing the filaments is altered.[16-19] They pointed out that at pH's greater than 5 the filaments have a net negative charge and there will be a strong electrostatic repulsive force tending to swell the lattice. If that were all, of course, the filaments would move so far apart that they would effectively be in solution. But expansion will be hindered by constraints provided by structural elements of the myofibril: the M-line linking adjacent thick filaments together, the Z-line linking adjacent thin filaments and the attached cross-bridges in rigor muscle linking thick and thin filaments together (Fig.2). It has recently been appreciated that attached cross-bridges in particular strongly inhibit changes in the spacing of the filament lattice.[20] Thus the spacing the filament lattice takes up depends on a balance between repulsive electrostatic forces and the constraining forces exerted by the transverse structural elements. The swelling that occurs with salt could thus be due either to an increase in the electrostatic repulsive force, or to a diminution of the structural restraints, or to both. Hamm[1] showed that sodium acetate, unlike sodium chloride, was ineffective at causing water uptake; this suggests that chloride ions play an active role. Following Hamm's suggestion we may suppose that a fraction of the Cl^- ions bathing the filaments bind to them, increasing their net negative charge and causing lattice expansion (Fig.3).

Although this provides the basis of an explanation, it seems insufficient to explain the remarkable co-operativity of swelling

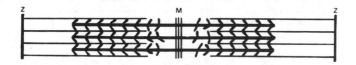

Figure 2 Structural elements of the myofibril restraining changes in filament lattice spacing. (Reproduced from Meat Science by permission).

and extraction with respect to chloride concentration. It is
likely, therefore, that in addition to chloride binding enhancing
the charge, salt reduces the structural restraints. Chloride
ions and pyrophosphate are known to have two effects which could
remove the cross-bridge restraint. Firstly, both chloride and
pyrophosphate tend to cause the dissociation of actomyosin. As
Bendall suggested,[5] pyrophosphate acts as an analogue of ATP;
it is a much less powerful dissociating agent than ATP, but its
properties as a dissociating agent are now well-documented.[21-24]
High concentrations of sodium chloride also tend to weaken the
actin myosin interaction.[25] Secondly, both sodium chloride and
pyrophosphate act to depolymerise thick filaments to myosin
molecules. This has been known in the case of sodium chloride
for a long time.[14] Less well-known is that pyrophosphate
enhances this effect.[26]

Thus, if in the presence of pyrophosphate the concentration
of sodium chloride is raised, it seems likely that the first
effect is to dissociate actomyosin, that is to detach
cross-bridges (Fig.4a). This removes the major restraint and
expansion of the filament lattice can occur. Depolymerisation of
the thick filament to free myosin molecules would be favoured
once the cross-bridges had dissociated.

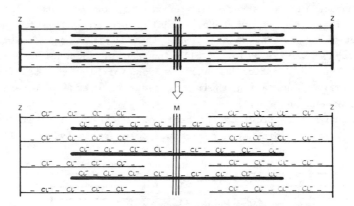

Figure 3 Diagram to illustrate how binding of some of the Cl⁻
ions to the filaments would increase their net negative charge
and cause swelling of the filament lattice. This mechanism is
insufficient to explain our data. (Reproduced from <u>Meat Science</u>
by permission).

In the absence of pyrophosphate a higher concentration of sodium chloride would be required for any effect to be apparent. The first effect, we suppose, would be to depolymerise the thick filament causing the thick filament shaft to disappear (Fig.4b). In the region where the thick filaments are overlapped by thin filaments the myosin molecules would remain attached to the thin filaments. Structural continuity between neighbouring thin filaments via the cross-bridges would thus be lost and expansion could again occur.

Although this explanation is speculative, it does appear to us to be reasonable and it does rest on well-established properties of myosin and actin and their behaviour with regard to salt and pyrophosphate. Future work will be required to test these ideas in detail.

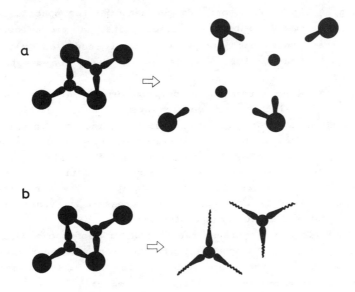

Figure 4 Alternative mechanisms for the removal of the cross-bridge constraint. The diagrams show end-on views of the filament lattice, with the large circles representing the thick filament shaft and the small circles the thin filaments. The pear-shaped objects represent cross-bridges. In (a) the cross-bridge constraint is removed by detachment of cross-bridges thus allowing expansion of the lattice. In (b) the cross-bridge constraint is removed by depolymerisation of the thick filament into myosin molecules which remain attached to the thin filaments. (Reproduced from <u>Meat Science</u> by permission).

LOSS OF WATER: DRIP

So far we have been discussing gains of water in meat. Can we explain losses by the reverse effect: the shrinking of myofibrils (Fig.1)? With regard to drip loss, when meat goes into rigor the myofibrils would be expected to shrink for two reasons. Firstly, as explained in previous chapters,[12,27] when the ATP becomes exhausted, the cross-bridges become attached. Goldman et al.[20] have shown using skinned fibres that at a sarcomere length of about 2.0 μm this attachment would cause the interfilament spacing to decrease by about 10%. Secondly, the pH fall from about 7 in the living muscle to about 5.5 in rigor would cause a reduction in negative charge on the filaments and reduce the electrostatic repulsive force. In glycerinated muscle such a pH fall causes a decrease also of about 10% in the lattice spacing.[17] Although no-one has yet measured the spacing change when intact muscle enters the rigor state, we might therefore expect that there should be a substantial reduction, although, for reasons that will be seen later, the reduction may be less than 20%.

How may the formation of drip be explained? Heffron and Hegarty[28] showed that when muscle entered the rigor state, the fibres shrank and there was a corresponding increase in the space between the fibres. This fibre shrinkage is presumably caused by the shrinkage of myofibrils which continue to be held together by desmin. Penny[29] supposed that the aqueous solution in the gaps between the fibres was the source of drip but there remains the question of what is the cause of the pressure that drives this solution out to the cut ends of the meat.

Plate 4a shows a transverse section of beef psoas muscle in the rigor state. In agreement with Heffron and Hegarty we can see gaps between the fibres but we are also able to see the endomysium revealed by the silver stain we use. It can be seen that a gap exists between the fibres and the enclosing endomysium; usually this gap surrounds most but not all of the fibre, but sometimes it completely surrounds it. This observation provides an important clue concerning the mechanism of drip formation.

We suppose that in the living muscle the collagen network, consisting of endomysium and perimysium, is stretched by the muscle fibres and is, therefore, under tension (Fig.5a). Thus the muscle fibres are compressed by the collagen network. When the fibre enters the rigor state it shrinks (Fig.5b). The collagen network, of course, continues to be in tension but the pressure it exerts will now be on the aqueous solution between the fibres rather than on the fibres themselves. This pressure will act to drive the aqueous solution along the gaps between the fibres and the endomysial sheaths to appear at the cut surfaces as drip.

Locker and Daines[30] showed that in cooked meat the cell membrane was disrupted; most of it remained attached to the endomysium but parts remained attached to the fibre. If this is also true of muscle in rigor, we can make no distinction between intracellular and extracellular space and we can see why the aqueous solution in the gap regions, and hence drip, should consist of a solution of sarcoplasmic proteins.[31]

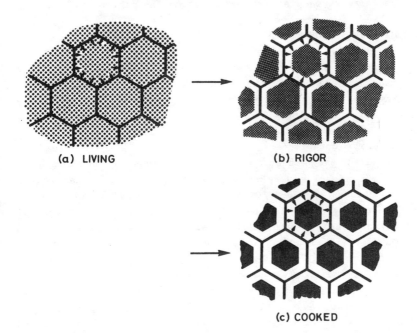

(a) LIVING   (b) RIGOR

(c) COOKED

Figure 5 Changes in water compartments of muscle occurring at rigor and after heating. The diagrams indicate the appearance of transverse sections of (a) pre-rigor muscle; (b) muscle in the rigor state; (c) cooked muscle. The arrows indicate the force exerted by the collagen network on the contents.

While this hypothesis needs to be stringently tested, there is a certain amount of circumstantial evidence to support it. We know that the endomysium is not a rigid tube since when in life muscle fibres contract, they do so at a constant volume and their diameters increase as they shorten.[32] Necessarily then the endomysium must be resilient and change diameter with length in a manner similar to the fibres. Endomysial sheaths can be prepared separated from their fibre contents and indeed have been shown to be highly elastic structures whose diameter at equilibrium appears to be less than that of the intact fibre.[33,34] Finally, it is known that when a fibre is skinned, that is the cell membrane and endomysial sheath are removed, the fibres swell substantially.[35,36] While a part of this effect may be osmotic due to the removal of sarcoplasmic proteins from the solution bathing the myofibrils, the effect is equally consistent with the collagen network mechanically compressing the fibres.

LOSS OF WATER: COOKING

On cooking, the losses of water from the cut ends of the meat are very great. This weight loss is accompanied by a shrinkage of the meat that occurs in two phases. At a temperature of about 45–60°C the shrinkage is primarily transverse to the fibre axis. At higher temperatures, 60–90°C, the shrinkage is primarily parallel to the fibre axis.[37]

Several authors[37-39] have simplified the problem of understanding the mechanism of cooking loss by showing that single fibres shrink very substantially in diameter when heated to temperatures of 50–60°C. (A discrepancy, however, exists between these groups with regard to whether or not longitudinal shortening occurs in single fibres at higher temperatures.)

The main question that concerns us here is whether the fibres actively shrink or are forced to shrink by the thermal shrinkage of the endomysial sheath surrounding the fibres (which is still present in the dissected fibres). We have answered this question in two ways. Firstly we have heated preparations of myofibrils (which of course do not contain collagen) at 55°C and pH 5.5 for 30 min (Plate 3). A dramatic reduction in diameter corresponding to a 40% reduction in volume occurs. This clearly demonstrates that myofibrils, and by implication muscle fibres within their endomysial sheaths, can actively shrink transversely. In preliminary experiments we have confirmed the results of Aronson[38] that myofibrils heated to higher temperatures (65°C) shrink longitudinally: this would account for the decrease in sarcomere length observed in cooked meat.[40,41]

Our other approach to the question is to heat, for 30 min, blocks of meat 1 x 1 x 3 cm with the long axis parallel to the fibres at a variety of temperatures from 40-90°C, either in a closed plastic bag, or by immersion in 0.85% saline. Both beef psoas and sternomandibularis muscles have been studied. After fixation in formol saline followed by standard wax embedding, transverse sections have been cut midway through the block. Our sections, which resemble those of Schmidt and Parrish[42] more than those by Locker and Daines[30], are shown in Plate 4. At 40°C there is a marked heterogeneity of the fibres: some have clearly shrunk considerably with a large gap between the fibres and the surrounding endomysium: others have not shrunk at all. This heterogeneity may be related to differences between fibre types.

At 50°C and 60°C a larger gap is present between nearly all the fibres and the surrounding endomysium. The cross-sectional area is about equally shared between fibres and the gaps between fibres. Although at 70°C and 80°C the meat has been treated at a temperature well above the melting point of collagen, the appearance is but little changed: the endomysium remains clearly visible and there are still large gaps between fibres and endomysium. At 90°C the endomysium, although detectable in places, is harder to see. Still, however, the gaps surrounding the fibres are present.

Such micrographs are very striking. After seeing them one cannot think of meat in quite the same way as before. The main point I want to make is that such micrographs provide no evidence that on cooking the collagen network is causing shrinking of the fibres by crushing them. When the collagen network thermally shrinks, it must squeeze on the aqueous solution in the gaps, not on the fibres themselves.

Longitudinal sections of beef psoas muscle cooked to a temperature of 70°C show the endomysium in the gap between adjacent fibres (Plate 5b). For long distances the endomysium and fibre are not in contact. Even after cooking at 90°C, the fibres remain essentially intact (Plate 5c).

These pictures suggest a simple and attractive model of the

cooking process. When the temperature of the meat reaches 40–60°C, the myofibrils, and hence the fibres, shrink, probably largely due to denaturation of myosin.[43] This greatly widens the gap, already present in rigor, between the fibres and their surrounding endomysium (Fig.5c). This effect alone would cause the flow of aqueous solution down the annular channels[44] to the cut ends greatly to be increased with concomitant transverse shrinking of the meat. When a temperature of 60–70°C is reached, denaturation of the collagen of the endomysium and perimysium occurs tending to shrink this network.[45–48] This will greatly increase the pressure on the aqueous solution causing it to be expelled more rapidly and the transverse shrinkage of the meat to become more rapid.

Both the endomysium and myofibrils appear to have the capacity to shrink longitudinally at temperatures of around 70°C. It is not yet clear which is the dominating influence and how the longitudinal shrinkage of the two is coupled.

It will be appreciated that the extent of shrinkage seen in a transverse section would be expected to depend on how far the section is made from the cut ends. We might expect that after the onset of rigor, or after a rapid temperature rise, the fibres would shrink to their new volume relatively rapidly. Immediately after such a shrinkage, we might expect the endomysium to exert equal pressures along its length (Fig.6a). Therefore, initially we would expect there to be no pressure gradient, and therefore no flow of fluid, for the greater part of the length of the fibre. However an abrupt pressure step would be expected at the ends of the fibre, from the pressure exerted by the endomysium just inside the meat to atmospheric pressure just outside. Hence initially the endomysium near and at the cut ends would be expected preferentially to shrink. Some time later we would expect the shape of an endomysial sheath to appear like that shown in Fig.6b, with the gap between fibre and endomysium much smaller at the ends than near the middle. As time goes on, of course, the endomysium even in the middle of the meat will slowly shrink.

To test this supposition, we have compared transverse sections of blocks of cooked beef sternomandibularis muscle made through the middle of the block with those as close as possible to the cut ends (Plate 6). We find that in the sections

made through the centre of the block, as with psoas muscle, there
is a wide gap between fibres and endomysium and the fibres are in
general rather well separated (Plate 6a), whereas in sections
made near the cut ends the gaps are much narrower and the fibres
are closer to one another and often touching (Plate 6b). Such
appearances explain rather well why the centre of a joint is much
more juicy (to the extent that water may be squeezed out by
gentle pressure[44]) than at the outside. It would be surprising
if there were no correlation between the juiciness of a piece of
meat and the fraction of the total water that was located outside
the fibres. The model explains in a qualitative way why the
water lost depends greatly on the distance between the cut
surfaces of the meat.[44]

To summarise, most of the water in meat is held within the
myofibrils in the narrow channels between the filaments.
Myofibrils are not inert: rather there is good evidence that
water uptake, such as occurs in meat processing, is caused by the

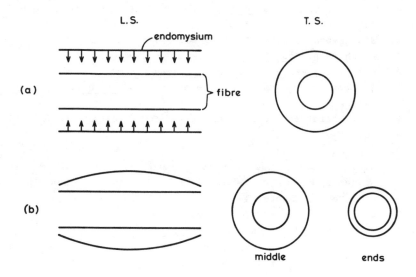

Figure 6 Predicted changes in the shape of an endomysial sheath
with time. On the left hand side of the diagram in (a) the
endomysium is shown in longitudinal section immediately after the
fibre has shrunk, while in (b) it is shown some time later when
its ends have collapsed. On the right hand side of the diagram
the corresponding transverse sections through the middle and
close to the cut ends are shown.

swelling of myofibrils. Similarly, losses, such as occur in drip or on cooking, are caused by the shrinking of myofibrils, the endomysium and perimysium acting to force the sarcoplasmic fluid expelled from the fibres along the annular gaps between fibres and their endomysial sheaths to the cut ends of the meat.

ACKNOWLEDGEMENT

We thank Mr A. Cousins for his excellent technical assistance in the preparation of the histological sections, and Professor A.J. Bailey, Mr P.D. Jolley, Dr P.J. Knight, Dr N.D. Light, Dr I.F. Penny and Mr A.A. Taylor for stimulating discussions.

REFERENCES

1. R. Hamm, Adv. Food Res., 1960, 10, 355.
2. R. Hamm, "Meat", Butterworths, London, 1975, p.321.
3. R. Hamm, "Muscle as a Food", Academic Press, USA, 1983, Vol. II.
4. M.D. Ranken, Chemistry & Industry, 1976, (18th December), 1052.
5. J.R. Bendall, J.Sci.Food Agric., 1954, 5, 468.
6. P. Sherman, Food Technol., 1961, 15, 79.
7. N.A. Iles, BFMIRA Sci.& Tech.Survey, 81, 1.
8. R. Hamm and R. Neraal, Z.Lebens.Unters.Forsch., 1977, 164, 243.
9. J.J. Macfarlane, G.R. Schmidt and R.H. Turner, J.Food Sci., 1977, 42, 1603.
10. G.R. Schmidt, Chapter 12.
11. H.E. Huxley, "The Structure and Function of Muscle", Academic Press, New York, 1972, Vol.I, 2nd Edn, p.301.
12. C.L. Davey, Chapter 1.
13. B.R. Eisenberg and A.M. Kuda, J.Ult.Res., 1976, 54, 76.
14. J. Hanson and H.E. Huxley, Symp.Soc.Exp.Biol., 1955, 9, 228.
15. G. Offer and J. Trinick, Meat Sci., 1983, 8, 245.
16. E. Rome, J.Mol.Biol., 1967, 27, 591.
17. E. Rome, J.Mol.Biol., 1968, 37, 331.
18. E. Rome, J.Mol.Biol., 1972, 65, 331.
19. G.F. Elliott, J.Theoret.Biol., 1968, 27, 71.
20. Y.E. Goldman, I. Matsubara and R.M. Simmons, J.Physiol., 1979, 295, 8OP.

21. W. Hasselbach, Biochim.Biophys.Acta, 1956, 20, 355.
22. A. Martonosi, M.A. Gouvea and J. Gergely, J.Biol.Chem., 1960, 235, 3169.
23. D. Gränicher and H. Portzehl, Biochim.Biophys.Acta, 1964, 86, 567.
24. L.B. Nanninga, Biochim.Biophys.Acta, 1964, 82, 507.
25. L.E. Greene, Biochemistry, 1981, 20, 2120.
26. W.F. Harrington and S. Himmelfarb, Biochemistry, 1972, 11, 2945.
27. R.E. Jeacocke, Chapter 3.
28. J.J.A. Heffron and P.V.J. Hegarty, Comp.Biochem.Physiol., 1974, 49A, 43.
29. I.F. Penny, J.Sci.Food Agric., 1975, 26, 1593.
30. R.H. Locker and G.J. Daines, J.Sci.Food Agric., 1974, 25, 1411.
31. A. Howard, R.A. Lawrie and C.A. Lee, Division of Food Preservation and Transport Technical Paper No.15, CSIRO, Australia, 1960.
32. G.F. Elliott, J. Lowy and C.R. Worthington, J.Mol.Biol., 1963, 6, 295.
33. H. Wang, Exp.Cell Res., 1956, 11, 452.
34. R.W. Fields, Biophys.J., 1970, 10, 462.
35. I. Matsubara and G.F. Elliott, J.Mol.Biol., 1972, 72, 657.
36. A. Magid and M.K. Reedy, Biophys.J., 1980, 30, 27.
37. R.L. Hostetler and W.A. Landmann, J.Food Sci., 1968, 33, 468.
38. J.F. Aronson, J.Cell Biol., 1966, 30, 453.
39. J.R. Bendall and D.J. Restall, Meat Sci., 1983, 8, 93.
40. B.G. Giles, Proc.15th Meeting Eur.Meat Res.Workers, 1969, p.289.
41. P.V.J. Hegarty and C.E. Allen, J.Food Sci., 1975, 40, 24.
42. J.G. Schmidt and F.C. Parrish, J.Food Sci., 1971, 36, 110.
43. I.F. Penny, J.Food Technol., 1967, 2, 325.
44. R.H. Locker and G.J. Daines, J.Sci.Food Agric., 1974, 25, 939.
45. V. Mohr and J.R. Bendall, Nature, 1969, 223, 404.
46. T.J. Sims and A.J. Bailey, "Developments in Meat Science - 2", Applied Science Publishers Ltd, Barking, Essex, 1981, p.29.
47. A.J. Bailey, Chapter 2.

48.   N.D. Light and D.J. Restall, unpublished results.

6
Recent Advances in the Chemistry of Rancidity of Fats

By Edwin N. Frankel

NORTHERN REGIONAL RESEARCH CENTER, AGRICULTURAL RESEARCH SERVICE,
U.S. DEPARTMENT OF AGRICULTURE, PEORIA, ILLINOIS 61604, U.S.A.

INTRODUCTION

There is a worldwide renewed interest in the problems of lipid oxidation as they relate to food deterioration and biological damage resulting from changes in diet, environment and ageing. Lipid peroxidation produces oxidative rancidity in foods and can decrease their nutritional quality and safety. There is now evidence that singlet oxygen plays a role in initiating autoxidation in unsaturated fats.[1-3] Singlet oxygen and other species of activated oxygen are implicated in many biological oxidation processes, in the universally important problem of ageing, as well as in environmental pollution.[4,5] This aspect of oxidation chemistry has become an intensely active area of research and much controversy has resulted.

This review deals with recent structural studies of primary and secondary lipid oxidation products, their volatile decomposition products and some of their flavour implications. Despite the considerable information available on the chemistry of lipid oxidation, the mechanism of flavour deterioration is not well understood. Studies on the source of volatile lipid oxidation products are still controversial and difficult to interpret.[6] A better understanding of the origin of these volatile oxidation products may provide the means for improving flavour stability of lipid-containing foods.

PRIMARY OXIDATION PRODUCTS

Free radical autoxidation

In the presence of initiators such as heat, metals or light, unsaturated lipids (LH) lose a hydrogen radical to form free radicals (L·) which rapidly react with oxygen to form peroxy radicals (LOO·) (Fig.1). The radical chain is propagated by

reacting the peroxy radical with more unsaturated lipids to form
hydroperoxides (LOOH) as the primary products of autoxidation.
Antioxidants can break this radical chain by reacting with LOO·
to form stable radicals that are too unreactive to propagate the
chain.

The decomposition of lipid hydroperoxides involves a very
complex set of reactions. Volatile breakdown products are formed
by homolytic cleavage abstracting hydroxy radicals and forming
alkoxy radicals (LO·). These breakdown products causing
rancidity in foods include complex mixtures of aldehydes,
ketones, alcohols, hydrocarbons, esters, furans and lactones.
Lipid hydroperoxides can also condense into dimers and polymers.
These high molecular weight materials can in turn oxidise and
decompose into volatile breakdown products. Further secondary
products consist of oxygenated monomeric materials including
epoxyhydroperoxides, ketohydroperoxides, dihydroperoxides, cyclic
peroxides and bicyclic endoperoxides. These secondary oxidation
products can also break down to produce volatile materials and
dialdehydes contributing to flavour degradation of foods.

The interactions of lipid hydroperoxides with proteins,

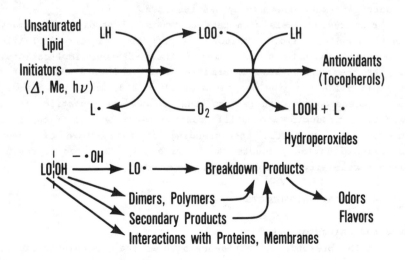

Figure 1 Free radical autoxidation of unsaturated lipids (LH)
and its consequences.

membranes and biological components are of most concern to
biochemists and food chemists. These reactions can affect vital
cell functions by damage to membranes, enzymes and proteins.
Fluorescent products are formed by the interaction of oxidised
lipids with amino acids, esters and amines by the formation of
conjugated Schiff bases. Fluorescent polymers are also formed by
the interaction of dialdehydes such as malonaldehyde with DNA.[7]
The fluorescence produced is related to loss of template activity
arising from disruption of the hydrogen-bonding system of DNA.
Lipid oxidation has also been related to membrane deterioration
caused by free radical mediated reactions that contribute to the
ageing process. Age pigments referred to as **lipofuscin** are
also fluorescent, and their formation is delayed by
administration of vitamin E. Fungal pigments related to age
pigments of higher cells have been attributed recently to
polymerised malonaldehyde derived from lipid peroxidation.[8] In
complex food systems, the interactions of lipid hydroperoxides
and their secondary products with proteins and other components
have a considerable impact on flavour stability and texture
changes during processing, cooking and storage. Some aspects of
these interactions will be briefly discussed in the last section
of this review. This subject has also been reviewed by
others.[9-12]

   The basic unsaturated fatty esters oleate, linoleate and
linolenate are present in all foods containing either vegetable
or animal fats and are susceptible to oxidation. Arachidonic
acid is present in the lipids of animal tissues at low levels,
and thus occurs in a variety of meats as well as in eggs. The
lipid extract from lean beef meat contains 2-4% triglycerides and
0.8-1% phospholipids.[13] The phospholipids include a large
proportion of polyunsaturated fatty acids (25% with 2 and 3
double bonds, 19% with 4 or more double bonds). Structural
studies of the primary hydroperoxide products of oleate,
linoleate, linolenate and arachidonate are therefore basic to an
understanding of the volatile decomposition materials causing
rancidity in foods.

   Autoxidation of oleate. The accepted mechanism involves
hydrogen abstraction on allylic carbons-8 and -11 to produce two
three-carbon allylic radicals. Oxygen attack at the end-carbon

positions of these intermediates produces a mixture of four allylic hydroperoxides with oxygen on carbons-8, -9, -10 and -11 positions (Fig.2). According to this mechanism, the isomeric allylic hydroperoxides would be expected to form in equal amounts. However, recent studies showed higher amounts of the 8- and 11-isomers than of the 9- and 10-isomers.[14],[15] Stereo-chemical studies showed a mixture of eight cis and trans allylic hydroperoxides.[16] With increase in temperature, the relative proportion of cis-8- and cis-11-isomers decreased, the trans-8- and trans-11-isomers as well as the cis-9- and cis-10-isomers increased, while the corresponding trans-9- and trans-10-isomers showed no change.

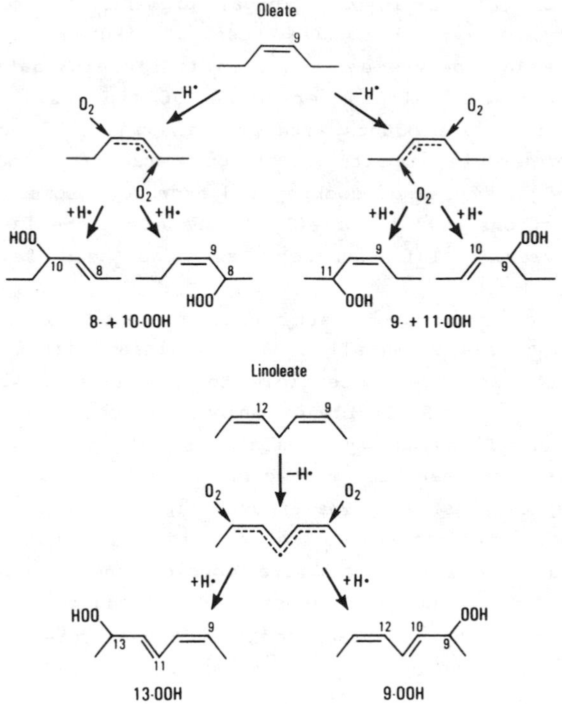

Figure 2 Mechanism of autoxidation of methyl oleate and linoleate.

   Autoxidation of linoleate. Hydrogen abstraction at the
doubly allylic carbon-11 position of linoleate produces a
pentadienyl radical (Fig.2). Reaction of this intermediate with
oxygen at the end-carbon positions produces a mixture of
conjugated diene 9- and 13-hydroperoxides. The formation of
equal proportions of these two isomers was shown to take place at
a wide range of oxidation temperatures.[17] Stereochemical studies
showed a mixture of four cis,trans- and trans,trans-conjugated
diene hydroperoxides and their relative proportions vary with

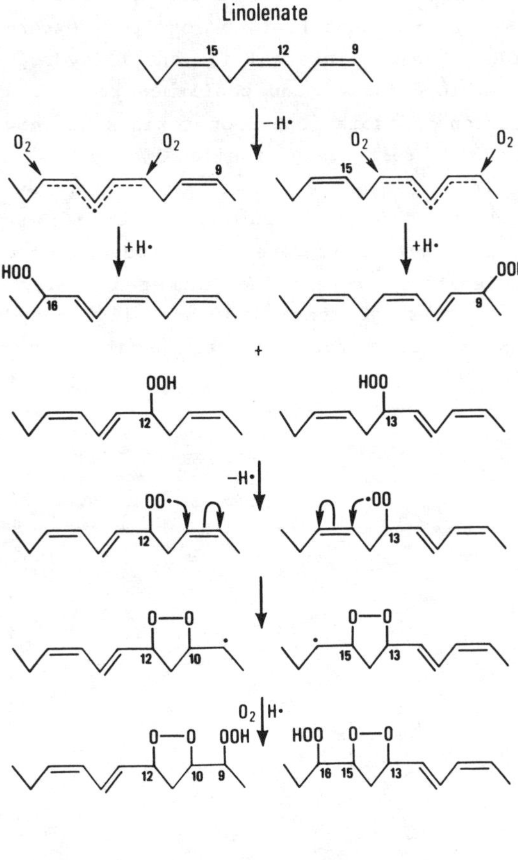

Figure 3   Mechanism of autoxidation of methyl linolenate and
cyclisation of internal 12- and 13-linolenate hydroperoxides.

substrate   concentration,   temperature   and   degree   of
oxidation.[17-20]   The autoxidation of dilinoleyl phosphatidyl-
choline in an aqueous suspension produced the same mixture of
conjugated diene 9- and 13-hydroperoxides as the autoxidation of
neat methyl linoleate.[21]

Autoxidation of linolenate.   Hydrogen abstraction at doubly
allylic carbon-11 and carbon-14 of linolenate produces two
pentadienyl radicals (Fig.3).   Reaction with oxygen at the
end-carbon positions produces a mixture of four conjugated diene
hydroperoxides containing a third isolated double bond.   The
significantly greater concentration of the external 9- and
16-hydroperoxides than the internal 12- and 13-hydroperoxides has
been known for a long time[22] and confirmed recently.[23,24]   This
uneven distribution of isomeric hydroperoxides has now been shown
to be due to the relatively facile 1,3-cyclisation of the
internal 12- and 13-hydroperoxides into hydroperoxy cyclic
peroxides   (Fig.3).[24-27]   Bicycloendoperoxides   structurally
related to the prostaglandins were also produced by autoxidation
of the 13-linolenate hydroperoxide isomer prepared by lipoxygenase action.[28]   This cyclisation of homoallylic hydroperoxides
apparently accounts also for the very small tendency of the

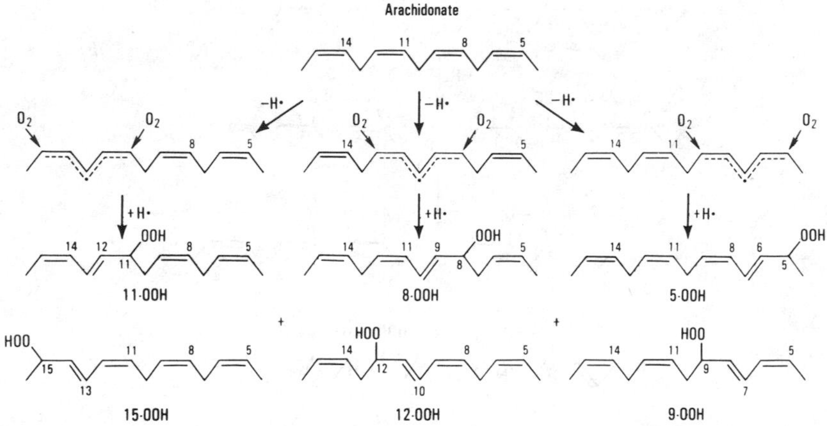

Figure 4 Mechanism of autoxidation of methyl arachidonate.

cis,trans-linolenate hydroperoxides to isomerise into the trans, trans-configuration.[23,29] Hydrogen-atom donors such as p-methoxyphenol[20] and tocopherol in large concentrations (5%)[30] also inhibit cyclisation and geometric isomerisation of linoleate and linolenate hydroperoxides.

Autoxidation of arachidonate. Hydrogen abstraction at doubly allylic carbon-7, -10 and -13 positions of arachidonate, produces three pentadienyl radicals reacting with oxygen at the end positions C-5 and C-9, C-8 and C-12, C-11 and C-15 (Fig.4). As in linolenate, the external 5- and 15-hydroperoxide isomers are formed in higher concentrations than the internal 8-, 9-, 11- and 12-OOH isomers.[31,32] The homoallylic structure of the internal hydroperoxides would allow cyclisation and explains their relatively low concentration compared to the external 5- and 15-hydroperoxide isomers.

Photosensitised oxidation

By exposure to light energy, a sensitiser such as chlorophyll becomes excited. In the presence of oxygen, the energy is transferred and oxygen from the triplet ground state becomes excited into the singlet state. The resulting singlet oxygen ($^1O_2$) is a very active species and reacts with linoleate at least 1500 times faster than normal triplet oxygen to form hydroperoxides (Fig.5).[1] Natural quenchers such as carotenoids protect lipids against photosensitised oxidation by interfering with this process.[4,33] In many foods, carotenoids are bleached out for cosmetic reasons. It may be wise to restore some of the carotenoids to the food for protection against $^1O_2$. However, this is not a simple solution because carotenoids are also autoxidised and break down into secondary products that can initiate and promote free radical oxidation, e.g. palm oil oxidation.

Photosensitised oxidation is not a free radical process and proceeds by a different mechanism to autoxidation. Singlet oxygen reacts directly with carbon-carbon double bonds by a concerted 'ene' addition. The resulting hydroperoxides are formed by oxygen addition at each unsaturated carbon. Oleate thus forms two hydroperoxide isomers, the 9- and 10-OOH with allylic trans unsaturation. Linoleate forms four hydroperoxide

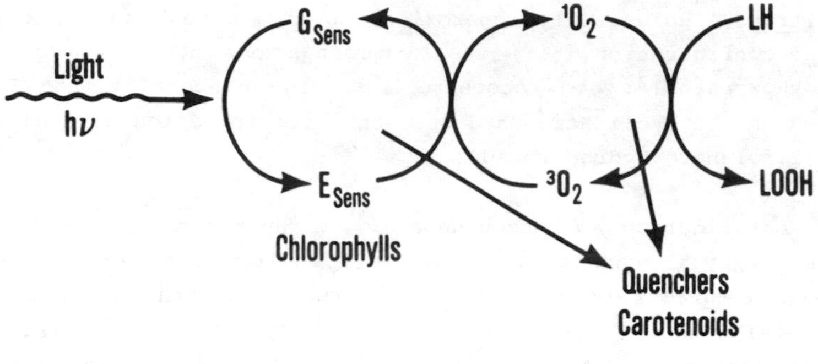

Figure 5 Generalised scheme for photosensitised oxidation of unsaturated lipids (LH).

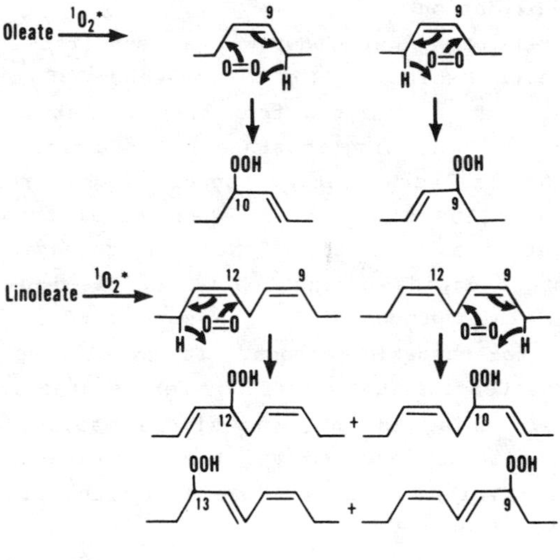

Figure 6 Mechanism of photosensitised oxidation of oleate and linoleate.[34]

isomers, two conjugated dienes (9- and 13-OOH) and two unconjugated dienes (10- and 12-OOH) (Fig.6). Similarly, linolenate produces six hydroperoxide isomers (9-, 10-, 12-, 13-, 15- and 16-OOH).[33],[35-37] According to the ene addition mechanism, one would expect an even distribution of isomeric hydroperoxides in all unsaturated fatty acids. However, our results show that the distribution is uneven in linoleate and linolenate.

The relative distribution of isomeric hydroperoxides produced by autoxidation and photosensitised oxidation are summarised in Fig.7. The lower relative concentration of the internal 12- and 13-hydroperoxides of autoxidised linolenate is due to their facile 1,3-cyclisation into hydroperoxy cyclic peroxides (Fig.3). After photosensitised oxidation, the lower relative concentration of the internal 10- and 12-hydroperoxides in linoleate and 10-, 12-, 13- and 15-hydroperoxides in linolenate is similarly due to their cyclisation into hydroperoxy cyclic peroxides.[38],[39] These internal hydroperoxide isomers have a cis double bond homoallylic to the hydroperoxide which permits

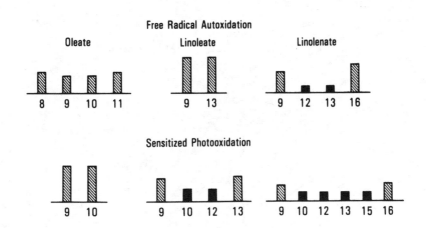

Figure 7 Distribution of isomeric hydroperoxides from free radical autoxidation and sensitised photooxidation of oleate, linoleate and linolenate.

a facile 1,3-cyclisation in linoleate (Fig.8)[38],[40] and linolenate.[39] In addition, hydroperoxy bis-cyclic peroxides and hydroperoxy bicycloendoperoxides were identified in photo-sensitised oxidised methyl linolenate.[39] These cyclisations are secondary free radical reactions in which methylene-interrupted unsaturated hydroperoxides apparently can lose a hydrogen radical even during oxidation with $^1O_2$.

SECONDARY OXIDATION PRODUCTS

Free radical autoxidation

The secondary products identified in highly autoxidised oleate included allylic ketoenes, saturated epoxy esters, dihydroxyenes and saturated dihydroxy esters.[14],[41] A multitude of compounds is found in thermally oxidised linoleate and decomposed linoleate hydroperoxides, including di- and trihydroxy esters derived from the corresponding keto- and hydroperoxy epoxides, ketodienes, epoxy hydroxy monoenes, dihydroxy and trihydroxy esters.[41],[42] Mechanistic pathways explaining the origin of these secondary oxidation products have been discussed

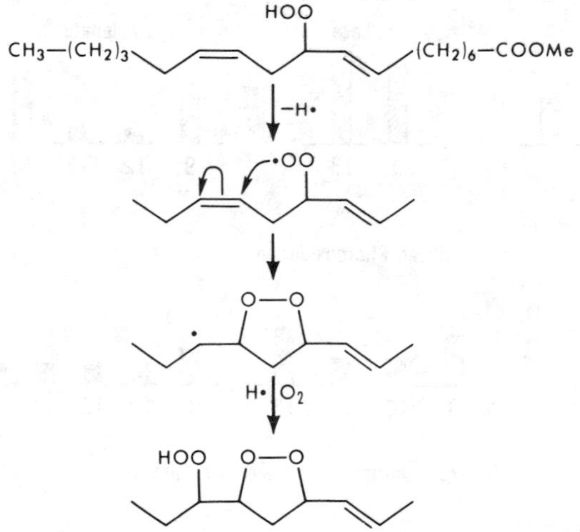

Figure 8 Cyclisation of 10-linoleate hydroperoxide from photosensitised oxidised methyl linoleate.

previously.[34,43]   The most important secondary products from autoxidised linolenate included hydroperoxy cyclic peroxides and dihydroperoxides.[25]   The hydroperoxy cyclic peroxides were shown to be derived by 1,3-cyclisation of the 12- and 13-linolenate hydroperoxides (Fig.3).   The dihydroperoxides are those expected from the secondary oxidation of the 9- and 16-linolenate hydroperoxides (Fig.9).   Hydroperoxy bicycloendoperoxides have also been prepared by a free radical initiated autoxidation of the 15-hydroperoxide isomer of arachidonic acid.[44]   Only recently has evidence accumulated that monomeric secondary oxidation products can undergo thermal decomposition into volatile products associated with flavour deterioration of fats.

The formation of oxidative dimers has also been demonstrated during the autoxidation of methyl linoleate.[45]   Peroxidic dimers linked by -C-O-O-C- bonds were formed at room temperature in slightly lower concentration than monohydroperoxides.   These dimeric products of autoxidation had functional groups that were similar to those of the monomeric hydroperoxides, including allylic OOH and allylic OH.[46]   Degradation of similar high molecular weight materials was previously shown to produce volatile aldehydes[47,48] and other compounds contributing to the flavour deterioration occurring in soybean oil.   Thermal polymerisation of oxidised fats during processing at elevated temperatures, such as in the deodourisation step (210-230°C),

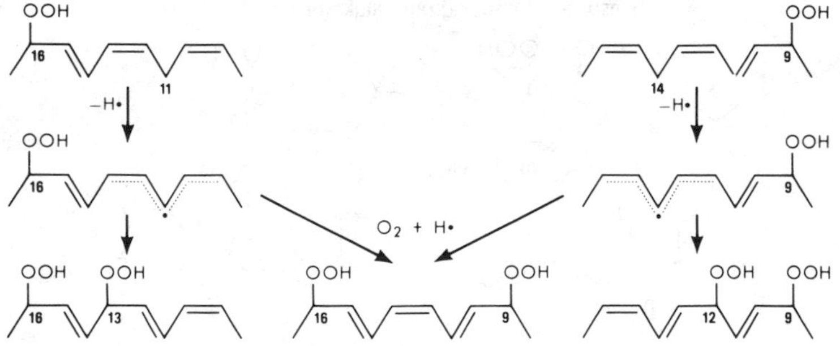

Figure 9   Formation of dihydroperoxides from 9- and 10-linolenate hydroperoxides.[25]

generates high molecular weight materials consisting mainly of dimers.[49] These non-volatile polymeric materials were associated with flavour and oxidative instability of soybean oil.[50,51]

Photosensitised oxidation

Hydroperoxy cyclic peroxides were identified as the major secondary products and dihydroperoxides as well as epoxy esters as minor secondary products in photosensitised oxidised methyl linoleate.[38] Thermal decomposition of the pure cyclic hydroperoxides produced some of the same volatile aldehydes and aldehyde esters as the monohydroperoxides. From photosensitised oxidised methyl linolenate, the major secondary products were also hydroperoxy cyclic peroxides and dihydroperoxides.[39] Minor products included keto- and epoxy-dienes, hydroperoxy bicyclic and bis-cyclic peroxides. Six-membered hydroperoxy cyclic peroxides have also been produced by the photosensitised oxidation of methyl linoleate hydroperoxides.[52] General structures for different hydroperoxy cyclic peroxides are summarised in Fig.10. The dihydroperoxides and cyclic peroxides

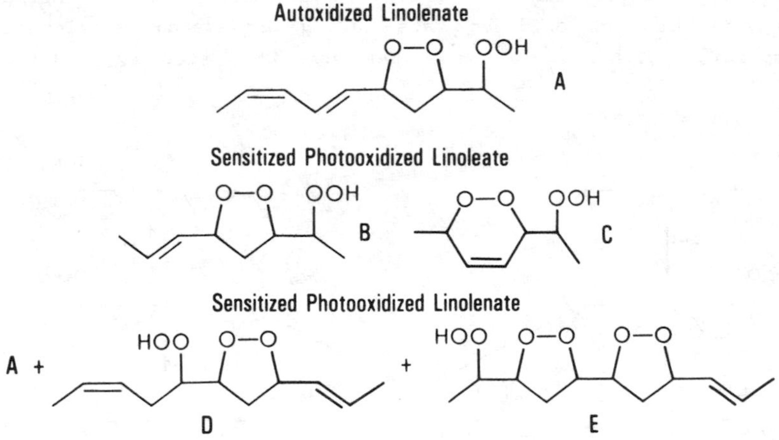

Figure 10 Structures of different hydroperoxy cyclic peroxides identified in autoxidised linolenate, sensitised photooxidised linoleate and linolenate.

have recently been shown to be rich sources of volatile materials produced by thermal decomposition. They would therefore be expected to serve as precursors of volatile flavour compounds.

VOLATILE DECOMPOSITION PRODUCTS

A generally accepted scheme for the fragmentation of monohydroperoxides involves the homolytic cleavage of the peroxy group to yield an alkoxy (LO·) and an ·OH radical (Fig.1). The next step involves the carbon–carbon cleavage on either side of the alkoxy radical to produce two types of aldehydes, an olefin radical and an alkyl radical (Fig.11). This reaction is known as β-scission in free radical chemistry. The radical products can, in turn, react with either ·OH or ·H. One vinyl alcohol derived from the reaction with ·OH will be unstable and tautomerise to a saturated aldehyde. The other alcohol ester would tend to polyesterify or form an estolide. The products from the reaction of the radical fragments with H· will be either an α-olefin, a hydrocarbon or a shorter chain

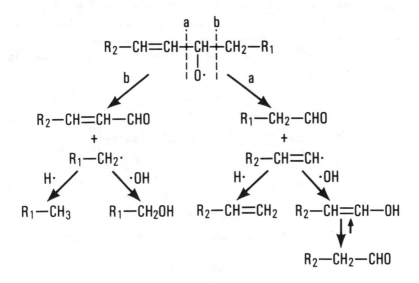

Figure 11 β-Scission of allylic secondary alkoxy radical from unsaturated fatty esters.

fatty ester. Most of the cleavage products expected by reactions in Fig.11 from oleate, linoleate and linolenate hydroperoxides have been identified in the corresponding oxidised fatty esters.[6]

Fat hydroperoxides

Much literature has appeared on the volatile products formed from oxidised fats, but only recently have these 'volatiles' been directly related to the hydroperoxide precursors. Using one of the simplest precursors, Evans et al.[53] were the first to decompose a single hydroperoxide prepared by lipoxygenase action. They thermally decomposed the 13-hydroperoxide isomer of linoleic acid by direct injection at 260°C onto a gas chromatograph and identified pentane as the principal hydrocarbon. The same approach was used in recent studies with the pure isomeric hydroperoxides of methyl linoleate[54] and with the mixture of isomers produced by autoxidation and photosensitised oxidation in methyl oleate, linoleate and linolenate.[55] Since the same major volatile cleavage products were obtained from either the 9- or 13-hydroperoxide isomer of linoleate, Chan et al.[54] suggested a carbon-oxygen scission in their decomposition, a path that is also supported for positional isomerisation of the hydroperoxides (Fig.12). However, this same C-O cleavage may also generate radical species capable of forming

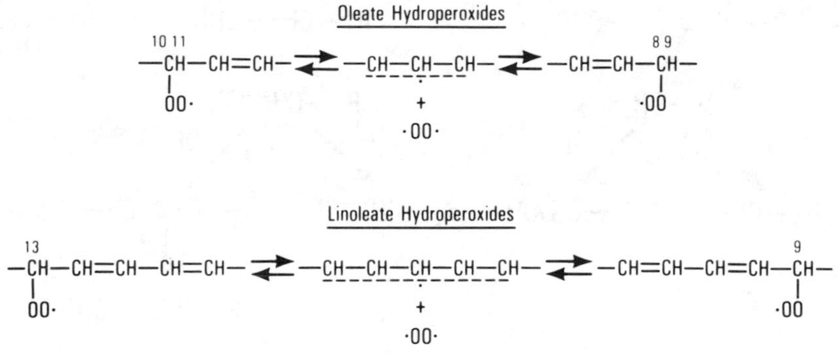

Figure 12   Positional rearrangement of oleate and linoleate hydroperoxides.

volatile products directly and not necessarily through an isomerised hydroperoxide. The composition of volatile fragmentation products will be controlled by the relative competitive rates between C-C cleavage a and b (Fig.11) and C-O cleavage during positional isomerisation (Fig.12).

All the aldehydes expected from the fragmentations a and b (Fig.11) have been found in studies with thermally decomposed hydroperoxides of methyl oleate, linoleate and linolenate. Major aldehydes produced by hydroperoxides from autoxidised oleate include 2-undecenal, 2-decenal, octanal, nonanal and decanal (Table 1). Aldehyde esters, hydrocarbons, fatty esters and alcohols expected from decomposition pathways a and b in Fig.11 have also been identified.[55] Shorter chain aldehydes (heptanal and 2-nonenal) than those predicted by these reactions can arise from bis-hydroperoxides[56] and by further oxidation of unsaturated aldehydes (discussed below). Thermal decomposition of the 9- and 10-hydroperoxides from photosensitised oxidised methyl oleate produced the expected volatiles from a mixture of 8-, 9-, 10- and 11-isomers.[55] This study showed a significant isomerisation of the 9- and 10-hydroperoxides into a mixture of 8-, 9-, 10- and 11-hydroperoxides, similar to those formed by autoxidation (Fig.12).

With the 9- and 13-linoleate hydroperoxides produced by autoxidation, the main volatiles expected by fragmentation at a and b (Fig.11) are methyl octanoate, 2,4-decadienal, methyl 9-oxononanoate, hexanal and pentane (Table 2). 2-Pentylfuran and 2-octenal are not expected from schemes in Fig.11 and can be rationalised from the decomposition of 9-linoleate hydroperoxide.[6,57] Other sources suggested for 2-pentylfuran include γ-ketononanal derived from a 10-hydroperoxide isomer[58] and γ-2-nonenolactone.[59] Important volatiles produced by the hydroperoxides from photosensitised oxidised methyl linoleate include methyl octanoate, 9-oxononanoate and 10-oxo-8-decenoate and 2-heptenal. Although in our studies 2-heptenal was unique to the hydroperoxides from photosensitised oxidation, and apparently originated from the 12-hydroperoxide isomer,[38] this carbonyl was an important product reported in autoxidised linoleate and fats containing linoleate. 2-Heptenal may originate from the 12-hydroperoxide that was identified in many vegetable oil esters at low levels of oxidation.[33,60,61]

Table 1 Volatile thermal decomposition products from methyl
oleate hydroperoxides[a]

Hydroperoxides	Volatiles	Autoxidation[b]	Photo-sensitised oxidation[c]
8-OOH	Me heptanoate	1.5%	4.9%
	Decanal	3.9	2.0
	2-Undecenal	1.7	7.1
	Me 8-oxooctanoate	3.5	3.0
9-OOH	Nonanal	(7.5)[d]	(5.0)[d]
	Me octanoate	5.0	9.7
	2-Decenal	5.4	12
	Me 9-oxononanoate	(7.5)[d]	(5.5)
10-OOH	Octane	2.7	10
	1-Octanol	0.4	1.0
	Nonanal	(7.5)[d]	(5.0)[d]
	Me 9-oxononanoate	(7.5)[d]	(5.5)[d]
	Me 10-oxo-8-decenoate	3.4	5.0
11-OOH	Heptane	4.4	4.6
	1-Heptanol	0.4	0.4
	Octanal	11	3.8
	Me 10-oxodecanoate	12	1.7
	Me 11-oxo-9-undecenoate	5.8	4.6
	Other volatiles	8.9	9.2

[a] (Frankel et al.[55]).

[b] Isomeric composition before thermal decomposition, 8-OOH:27%; 9-OOH:23%; 10-OOH:23%; 11-OOH:27%

[c] Isomeric composition before and after thermal decomposition, 8-OOH:0, 18%; 9-OOH:50, 26%; 10-OOH:50, 31%; 11-OOH:0, 25%.

[d] Values were divided by assuming that volatile came equally from 9-OOH and 10-OOH.

Table 2 Volatile thermal decomposition products from methyl linoleate hydroperoxides[a]

Hydroperoxides	Volatiles	Autoxidation[b]	Photo-sensitised oxidation[c]
9-OOH	Me octanoate	15%	7.6%
	2/3-Nonenal	1.4	(0.8)[d]
	2,4-Decadienal	14	4.3
	Me 9-oxononanoate	19	(11)[d]
10-OOH	1-Octen-3-ol	Tr	1.9
	2-Nonenal	−	(0.8)[d]
	Me 9-oxononanoate	−	(11)[d]
	Me 10-oxo-8-decenoate	4.9	14
12-OOH	Hexanal	−	(6.5)[e]
	2-Heptenal	Tr	9.9
13-OOH	Pentane	9.9	4.3
	Pentanal	0.8	0.3
	1-Pentanol	1.3	0.3
	Hexanal	15	(6.5)[e]
	Other volatiles	18.7	20.8

[a] (Frankel *et al.*[55]).

[b] Isomeric composition before and after thermal decomposition, 9-OOH:50, 47%; 10-OOH:0, 2%; 12-OOH:0, 4%; 13-OOH:50, 47%.

[c] Isomeric composition before and after thermal decomposition, 9-OOH:32, 28%; 10-OOH:17, 19%; 12-OOH:17, 21%; 13-OOH:34, 32%.

[d] Values were divided by assuming that volatile came equally from 9-OOH and 10-OOH.

[e] Values were divided by assuming that volatile came equally from 12-OOH and 13-OOH.

Table 3 Volatile thermal decomposition products from methyl
linolenate hydroperoxides[a]

Hydroperoxides	Volatiles	Autoxidation[b]	Photo-sensitised oxidation[c]
9-OOH	Me octanoate	22%	15%
	3,6-Nonadienal	0.5	(0.55)[e]
	Decatrienal	14	4.8
	Me 9-oxononanoate	13	(6.0)
10-OOH	3,6-Nonadienal	–	(0.55)[e]
	Me 9-oxononanoate	–	(6.0) [d]
	Me 10-oxo-8-decenoate	4.2	13
12-OOH	2/3-Hexenal	(0.7)[d]	(1.7)[d]
	2,4-Heptadienal	9.3	8.8
13-OOH	2-Pentenal	1.6	1.2
	2/3-Hexenal	(0.7)[d]	(1.7)[d]
15-OOH	Propanal/Acrolein	–	(4.5)[f]
	2-Butenal	0.5	11
16-OOH	Ethane/Ethene	10	3.2
	Propanal/Acrolein	7.8	(4.5)[f]
	Other volatiles	15.7	17.5

a   (Frankel et al.[55]).

b   Isomeric composition before and after thermal decomposition,
9-OOH:32, 31%; 12-OOH:11, 10%; 13-OOH:11, 14%; 16-OOH:46, 45%.

c   Isomeric composition before and after thermal decomposition,
9-OOH:23, 22%; 10-OOH:13, 14%; 12-OOH:12, 8%; 13-OOH:14, 13%;
15-OOH:13, 15%; 16-OOH:25, 28%.

d   Values were divided by assuming that volatiles came equally
from 12-OOH and 13-OOH.

e   Values were divided by assuming that volatiles came equally
from 9-OOH and 10-OOH.

f   Values were divided by assuming that volatiles came equally
from 15-OOH and 16-OOH.

Major volatiles identified from linolenate hydroperoxides formed by autoxidation include methyl octanoate, decatrienal, methyl 9-oxononanoate, 2,4-heptadienal, propanal and acrolein (Table 3). The hydroperoxides produced by photosensitised oxidation generated more methyl 10-oxo-8-decenoate and 2-butenal. Many common products formed from both types of hydroperoxides (autoxidation and photosensitised oxidation) such as 3-hexenal, propanal and methyl 9-oxononanoate, also may be derived by thermal cleavage of the hydroperoxy cyclic peroxides formed as secondary products from the 12- and 13-linolenate hydroperoxides (Fig.3).

Although no report has been made on the decomposition of arachidonate hydroperoxides, the volatile products of autoxidised arachidonic acid may be expected to be similar to those of linoleate because both of these fatty acids have an n-6 double bond. Thus, the major volatile carbonyls identified in arachidonic acid autoxidised at 20°C indeed include hexanal and 2,4-decadienal.[62] Hexanal would be derived from the 15-arachidonate hydroperoxide isomer that has an n-6 hydroperoxide group just as in 13-linoleate hydroperoxide (Fig.13). Similarly, 2,4-decadienal would originate from the 11-arachidonate hydroperoxide that corresponds to the 9-hydroperoxide isomer of linoleate.

Secondary products

Many of the volatile products identified in oxidised lipids can be formed either by decomposition of monohydroperoxides or by further oxidation and breakdown of secondary products (Fig.1). The multitude of secondary products that can be formed by both autoxidation and photosensitised oxidation has already been discussed earlier. These secondary products can undergo fragmentation by similar pathways to those shown in Fig.11. Some of the breakdown products of secondary products from linoleate are shown in Fig.13. Epoxy hydroperoxides are the probable precursors of epoxy aldehydes identified in oxidised butterfat,[63] trilinolein[64] and from thermally decomposed methyl linolenate hydroperoxides.[55] Dihydroperoxides may produce lower molecular weight aldehydes and dialdehydes.[6] Oxygenated hydroperoxides are also potential precursors of the corresponding hydroxy and oxo aldehydes that have been reported as volatile

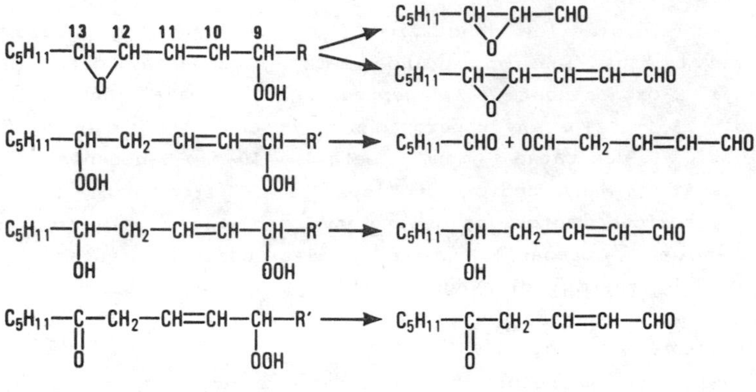

$$C_5H_{11}-\overset{13}{C}H-\overset{12}{C}H-\overset{11}{C}H=\overset{10}{C}H-\overset{9}{C}H-R$$

with the 13-position forming an epoxide (O bridge) and 9-position bearing OOH, giving:

$$C_5H_{11}-CH-CH-CHO \quad (\text{epoxide at 13,12})$$

$$C_5H_{11}-CH-CH-CH=CH-CHO \quad (\text{epoxide})$$

$$C_5H_{11}-\underset{OOH}{CH}-CH_2-CH=CH-\underset{OOH}{CH}-R' \longrightarrow C_5H_{11}-CHO + OCH-CH_2-CH=CH-CHO$$

$$C_5H_{11}-\underset{OH}{CH}-CH_2-CH=CH-\underset{OOH}{CH}-R' \longrightarrow C_5H_{11}-\underset{OH}{CH}-CH_2-CH=CH-CHO$$

$$C_5H_{11}-\underset{O}{\overset{\parallel}{C}}-CH_2-CH=CH-\underset{OOH}{CH}-R' \longrightarrow C_5H_{11}-\underset{O}{\overset{\parallel}{C}}-CH_2-CH=CH-CHO$$

Figure 13 Decomposition of secondary oxidation products of linoleate.[6]

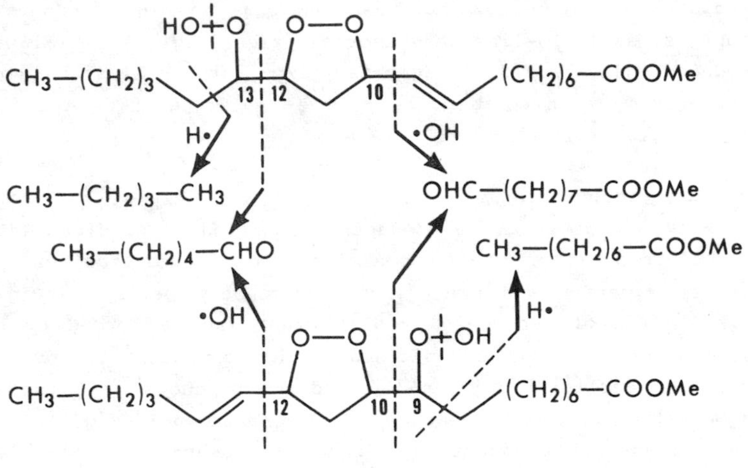

Figure 14 Thermal decomposition of hydroperoxy cyclic peroxides from photosensitised oxidised methyl linoleate.[38]

products of unsaturated fatty esters.[59]

Recent studies on the thermal decomposition of pure hydroperoxy cyclic peroxides have provided direct evidence that these secondary products may be important precursors of flavour volatile materials. The cyclic peroxides from methyl linoleate produced some of the common volatile materials identified from the hydroperoxides obtained by either autoxidation or photo-sensitised oxidation (Fig.14).[55] The 13-hydroperoxy cyclic peroxide isolated from photosensitised oxidised methyl linoleate produced hexanal and methyl 10-oxo-8-decenoate as major volatiles.[37] The 9-hydroperoxy cyclic peroxide generated 2-heptenal and methyl 9-oxononanoate. The same fragmentation paths as discussed for monohydroperoxides (Fig.11) also explain the formation of many of these volatile products. Other fragmentation pathways suggested by the product composition included cleavage of the peroxide ring and carbon-carbon bond β to the _trans_ allylic double bond. With hydroperoxy cyclic peroxides from autoxidised methyl linolenate, the main cleavage occurred between the peroxide ring and the carbon-bearing hydroperoxide group.[65] New volatiles identified included methyl 8-(2-furyl)-octanoate, methyl ketones and conjugated diunsaturated aldehyde esters.

The suggestion of further oxidation of unsaturated aldehydes as a source of additional volatile products[66] was later confirmed in several studies of the oxidation of unsaturated aldehydes.[6] The best evidence indicates that at low temperature mono- and di-unsaturated aldehydes oxidise readily and more rapidly than the unsaturated fatty ester precursors, whereas the saturated aldehydes are much less susceptible to oxidation.[57,67] During oxidation of unsaturated fats, saturated aldehydes would accumulate, whereas the unsaturated aldehydes oxidise further. The secondary products of unsaturated aldehydes include lower molecular weight aldehydes and dialdehydes including malonaldehyde. The mechanism postulated for oxidation involves α-hydroperoxidation producing lower saturated aldehydes and dialdehydes from 2-enals, keto-aldehydes and alcohols from 2,4-dienals (Fig.15). Other products include hydrocarbons, benzene, mono- and dibasic acids. More recently, oxidation of 2,4-decadienal with alkyl radicals was postulated to occur via a 3-carbon allylic radical undergoing oxygen attack on one end to

produce   2-octenal[57]   (Fig.16).   Although   2,4-decadienal   is
directly formed from the 9-hydroperoxide isomer of linoleate,
hexanal can originate from either the 13-hydroperoxide isomer or
from   further   oxidation   of   2,4-decadienal.   Therefore,   the
relative   proportion   of   these   two   volatiles   depends   on   the
autoxidation   conditions.   The   presence   of   oxygen   and   low
oxidation temperatures generally leads to a relatively higher
proportion of hexanal than 2,4-decadienal.   The   concept   of
preferential hydroperoxide scission a versus b (Fig.11) to
explain the effect of temperature on the relative ratio of these
aldehydes,[68],[69]   is   not   supported   by   the   recent   evidence[57],[70]
favouring competitive oxidation of aldehydes.

Flavour significance

Minor volatile oxidation products may assume an importance
out of all proportion to their relative concentration in foods
since they can be extremely potent and can affect flavour
properties at levels below 1 ppm.   There is also a considerable

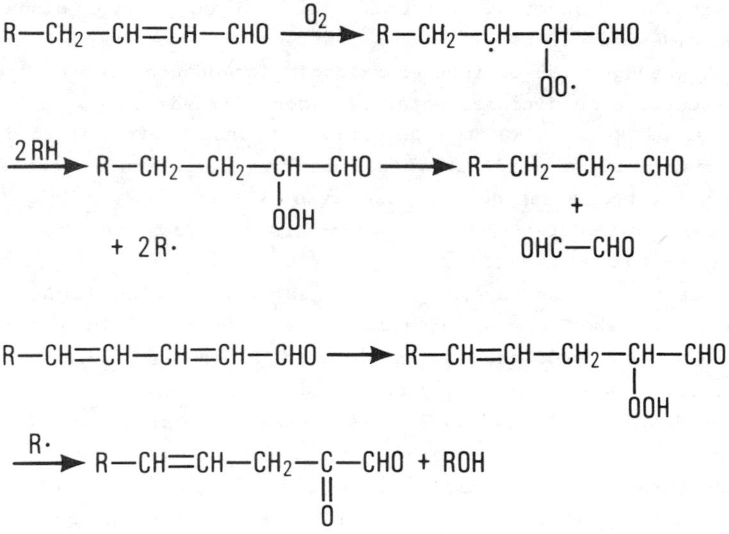

Figure 15 Oxidation of enals and dienals.[67]

difference in the flavour significance of volatile lipid oxidation compounds. The types of flavour developed from many of these compounds depend on a multitude of complex interactions, concentration ranges and the medium in which they are tasted. Hydrocarbons have the highest threshold value (90-2150 ppm) and presumably have the least impact on flavour. Substituted furans (2-27 ppm), vinyl alcohols (0.5-3 ppm) and l-alkenes (0.02-9 ppm) are not particularly significant. In increasing importance are 2-alkenals (0.04-2.5 ppm), alkanals (0.04-1.0 ppm), trans-2, trans-4-alkadienals (0.04-0.3 ppm), certain alkadienals (0.002-0.3 ppm), certain cis-alkenals (0.0003-0.1 ppm), trans-2,cis-4-alkadienals (0.002-0.06 ppm) and vinyl ketones (0.00002-0.007 ppm).

Table 4 lists volatile carbonyls identified in beef fat in decreasing order of relative concentration. If the threshold values are considered, the most abundant carbonyl, 2-decenal, is

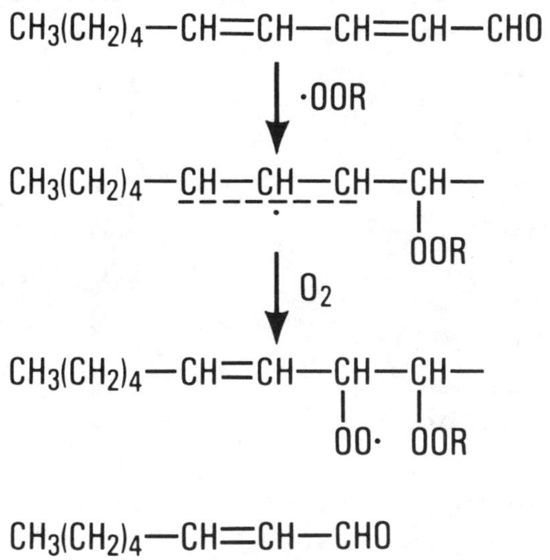

$$CH_3(CH_2)_4-CH{=}CH-CH{=}CH-CHO$$

$$\downarrow \cdot OOR$$

$$CH_3(CH_2)_4-CH{-}CH{-}CH{-}CH{-}$$
$$| \atop OOR$$

$$\downarrow O_2$$

$$CH_3(CH_2)_4-CH{=}CH-CH{-}CH{-}$$
$$OO\cdot \quad OOR$$

$$CH_3(CH_2)_4-CH{=}CH-CHO$$

Figure 16 Mechanism of 2-octenal formation from 2,4-decadienal.[57]

apparently not very flavour significant. The carbonyls formed in
smallest concentration, 2,6-nonadienals and 4-heptenals, would be
among the most significant. cis-4-Heptenal was also identified
from lipids in cod muscle as an extremely potent aroma compound
with a flavour threshold value of 0.0005-0.0016 ppm in oil.[71]
4-Heptenal was described as having a 'green' odour, characterised
in high dilution as tallowy, creamy and butterscotch-like, and
2,6-nonadienal was described as 'green'-cucumber-like and a
tallowy odour.[72] A large number of compounds have been identified in the aroma of meat products exposed to heat, but none has
been described as having a unique meaty odour.[73] The wide
diversity of flavour descriptions has led to controversy because
different terms are used by different people to characterise the
same flavour and odour. To evaluate the importance of volatile

Table 4 Carbonyls from beef fat

Carbonyls in order of concentration	Flavour threshold[a]	Flavour significance
1 2-Decenal[b]	5.5	12
2 2-Undecenal[b]	4.2	11
3 2-Nonenal[b]	0.1	5
4 2-Octenal[b]	1.0	10
5 Hexanal[b]	0.15	6
6 2-Heptenal[b]	0.63	9
7 Heptanal[b]	0.042	3
8 Nonanal[b]	0.32	8
9 Octanal[a]	0.068	4
10 2,4-Decadienal[b]	0.28	7
11 4 cis-Heptenal[c]	0.001[c]	1
12 4 trans-Heptenal[c]	0.32[c]	8
13 2 trans,6-trans Nonadienal[c]	0.001[c]	1
14 2 trans,6-cis Nonadienal[c]	0.002[c]	2

[a]Meijboom[74]   [b]Yamato et al.[75]   [c]Hoffman & Meijboom[72]

products individually, in combination and permutation under different conditions, may appear to be a staggering task. However, much progress has been made recently in methodology to identify and analyse extremely small quantities of flavour compounds[76]. With improved panel procedures, this apparently staggering problem may find a solution in the future.

INTERACTION PRODUCTS WITH FOOD COMPONENTS

Many natural non-lipid components in foods can influence lipid oxidation in different ways. The effects of natural phenolic antioxidants, synergists and trace metals are well known but their interactions with various food components may have a considerable impact on the development of rancidity. Depending on processing and storage conditions some food components may accelerate, retard or have no effect on lipid oxidation.

Water and its controlling effects on lipid oxidation during storage of foods have been known for a long time.[9,12,77] At very low levels, water has an antioxidant effect that has been explained in different ways, namely: by decreasing the prooxidant activity of metals (formation of metal salt hydrates which are less fat soluble), by preventing decomposition of lipid hydroperoxides (stabilisation by hydrogen-bonding with water), and by promoting nonenzymic browning (producing antioxidant materials). At high concentrations, water acts as a prooxidant by increasing mobilisation of reactants such as metallic catalysts dissolved in the water phase and allowing more intimate association with the lipid-water interface where hydroperoxides concentrate. Water has a profound effect on interactions between lipid oxidation products and other food components, especially haem compounds, amino acids and proteins. Water activity strongly affects cross-linking of proteins, enzyme inactivation, protein scission, production of protein radicals and destruction of amino acids.

Phospholipids and other surface-active compounds may act either as antioxidants or prooxidants. In a review of these activities for phospholipids,[78] the antioxidant effect was ascribed to the regeneration of phenolic inhibitors and to complexation of prooxidant metals. The metal chelating effect of phospholipids and especially phosphoinositols has been

demonstrated.[79,80] Other inhibiting effects include the tendency
of phospholipids to react with lipid hydroperoxides, to catalyse
their decomposition or form inert peroxidic polymers. On the
other hand, prooxidant effects of phospholipid in meat tissue
would be expected by oxidation of the polyunsaturated fatty acid
components and by interaction of the oxidation products with
other meat components such as carbohydrates and proteins. The
brown products formed by heating lecithin with water at 180 °C
have been attributed to polymers of lecithin due to aldol
condensation.[81] There is also recent evidence of interaction
between volatile oxidation products and the amino moiety of
phospholipids. When the volatile products were compared between
tripalmitin and 1,2-dipalmitoyl-glycerol-3-phosphoethanolamine
heated at 150 °C, alkanals were absent in the oxidised
phospholipid.[82]

Haem and **haematin** compounds are natural iron complexes
present in adipose tissue that have greater prooxidant activity
than any other known catalysts.[83] Catalytic activities for the
decomposition of linoleate hydroperoxides were listed in the
decreasing order: catalase, haemoglobin, haematin, cytochrome c
and peroxidase. Cytochrome P450, responsible for liver
microsomal lipid peroxidation, was more effective than other
haemoproteins and haematin as lipid oxidation catalysts.[84] A
mechanism in which the catalyst acts as a peroxidase was invoked
in the oxidative deterioration of lipid-containing foods. By
treating cytochrome P450 with detergents or proteases a high-spin
cytochrome P420 was produced with significantly increased
activity for linoleic acid hydroperoxide decomposition that
exceeded that of haematin and methaemoglobin.[85] ESR studies[86]
showed that cytochrome c can react with products of lipid
oxidation. The prooxidant effect of haem compounds is influenced
by the level of water in foods. In model systems containing fat
and casein, the effect of haem compounds was very little in the
absence of water and increased appreciably in a medium like meat
containing 60% water.[12] Myoglobin was similarly found to have no
prooxidant effect in dry turkey meat but was very active after
rehydration of the meat.[87]

In the presence of **proteins** and **amino acids,** the
development of fat rancidity is greatly modified because of their
interactions with lipid oxidation products. The chemistry of the

complex interactions between lipid hydroperoxides and secondary products with protein/amino acids has been reviewed extensively.[10],[11],[88] However, the impact of these reactions on flavour deterioration of food proteins is not well understood. Hydroperoxide radicals are very reactive with sulphur[89] and amine[90],[91] functional groups of amino acids and protein. Aldehydes and epoxides derived from the decomposition of lipid hydroperoxides also react with thiols from cysteine.[11] Imino Schiff bases are the principal products of the reaction of aldehydes and dialdehydes with amines.[7] Schiff bases can easily polymerise by aldol condensation producing dimers and complex high molecular weight brown materials known as **melanoidins** that have not been characterised. Changes of flavour are due to changes of volatile lipid oxidation products from aldolisation with amino groups of proteins.

Carbonyls forming Schiff bases are reported to lose their effect on rancid odour in foods.[92] But these Schiff bases and resulting high molecular weight brown materials are unstable and produce new volatiles that affect the flavour characteristics of foods, especially during cooking and processing. From a mixture of heptanal and tyrosine ester, 2-alkenals are produced and hydrolysis of the aldol condensation product was suggested.[93] Odour intensity of octanal stored with casein at 60° and 46% moisture decreased with storage time and produced glue-like off-flavours.[91] Changes in food texture and rheological properties are also associated with the polymerisation of the lipid and peptide chains. In general, the rancid flavour of oxidised fats is masked or modified by their reactions with proteins and other derivatives. When beef or chicken meat are fried, secondary lipid oxidation products react with meat proteins producing undesirable flavour development.[12] Fried meat undergoes further flavour deterioration on storage through the oxidised fat adhering to the food surface. Much research has recently appeared on the interaction of lipid oxidation products with selected amino acids and proteins. However, there is a dearth of information on the complex high molecular weight interaction products formed during actual cooking and frying of meat and how they are degraded into volatile compounds that affect flavour.

Browning Maillard reaction products from reducing sugars and amino acids also have a marked influence on lipid oxidation

and contribute to meat flavours.[94] These browning materials
generally retard lipid oxidation.[95-97] This antioxidant effect
may be due to the reducing properties of browning materials such
as reductones[98] which would be expected to stabilise
hydroperoxides by their reduction to the corresponding allylic
alcohols. This inhibiting effect has also been related to the
metal chelating property of browning materials because it is
increased in the presence of small concentrations of metal.[12]
Clearly, much more research is needed to clarify the complex
interactions of lipid oxidation products with various food
components, if we are to solve problems of cooking and storing
foods and meats where, at present, loss of flavour cannot be
avoided.

REFERENCES

1. H.R. Rawls and P.J. van Santen, J.Am.Oil Chem.Soc., 1970,
    47, 121.
2. A.M. Clements, R.M. van den Engh, D.J. Frost, K. Hoogenhout
    and J.R. Nooi, J.Am.Oil Chem.Soc., 1973, 50, 325.
3. D.J. Carlsson, T. Suprunchuk and D.M. Wiles, J.Am.Oil
    Chem.Soc., 1976, 53, 656.
4. C.S. Foote, "Free Radicals in Biology", Academic Press, New
    York, 1976, Vol.II, p.85.
5. N.I. Krinsky, Trends Biochem.Sci., 1977, 2, 35.
6. E.N. Frankel, Prog.Lipid Res., 1983, 22, 1.
7. A.L. Tappel, "Free Radicals in Biology", Academic Press, New
    York, 1980, Vol.IV, p.2.
8. K.D. Munkres, "Age Pigments", Elsevier/North-Holland
    Biomedical Press, 1981, p.83.
9. T.P. Labuza, CRC Crit.Rev.Food Technol., 1971, 2 (3), 355.
10. M. Karel, K. Schaich and R.B. Roy, J.Agric.Food Chem., 1975,
    23, 159.
11. H.W. Gardner, J.Agric.Food Chem., 1979, 27, 220.
12. J. Pokorny, Rev.Franc.Corps Gras, 1981, 28, 151.
13. I. Hornstein, P.F. Crowe and M.J. Heimberg, J.Food Sci.,
    1961, 26, 581.
14. E.N. Frankel, W.E. Neff, W.K. Rohwedder, B.P.S. Khambay,
    R.F. Garwood and B.C.L. Weedon, Lipids, 1977, 12, 901.
15. H.W.-S. Chan and G. Levett, Chem.Ind.,Lond., 1977, 692.
16. R.F. Garwood, B.P.S. Khambay, B.C.L. Weedon and E.N.

Frankel, J.Chem.Soc.Chem.Commun., 1977, 364.

17. E.N. Frankel, W.E. Neff, W.K. Rohwedder, B.P.S. Khambay, R.F. Garwood and B.C.L. Weedon, Lipids, 1977, 12, 908.

18. E.N. Frankel, "Fatty Acids", American Oil Chemists' Society, Champaign, Illinois, USA, 1979, p.353.

19. H.W.-S. Chan and G. Levett, Lipids, 1977, 12, 99.

20. N.A. Porter, B.A. Weber, H. Weenen and J.A. Khan, J.Am.Chem.Soc., 1980, 102, 5597.

21. C.G. Crawford, R.D. Plattner, D.J. Sessa and J.J. Rackis, Lipids, 1980, 15, 91.

22. E.N. Frankel, C.D. Evans, D.G. McConnell, E. Selke and H.J. Dutton, J.Org.Chem., 1961, 26, 4663.

23. H.W.-S. Chan and G. Levett, Lipids, 1977, 12, 837.

24. E.N. Frankel, W.E. Neff, W.K. Rohwedder, B.P.S. Khambay, R.F. Garwood and B.C.L. Weedon, Lipids, 1977, 12, 1055.

25. W.E. Neff, E.N. Frankel and D. Weisleder, Lipids, 1981, 16, 439.

26. D.T. Coxon, K.R. Price and H.W.-S. Chan, Chem.Phys.Lipids, 1981, 28, 365.

27. I. Toyoda, J. Terao and S. Matsushita, Lipids, 1982, 17, 84.

28. D.E. O'Connor, E.D. Mihelich and M.C. Coleman, J.Am. Chem.Soc., 1981, 103, 223.

29. N.A. Porter, L.S. Lehman, B.A. Weber and K.J. Smith, J.Am.Chem.Soc., 1981, 103, 6447.

30. K.E. Peers, D.T. Coxon and H.W.-S. Chan, J.Sci.Food Agric., 1981, 32, 898.

31. N.A. Porter, J. Logan and V. Kontoyiannidou, J.Org.Chem., 1979, 44, 3177.

32. J. Terao and S. Matsushita, J.Food Process.Preserv., 1980, 3, 329.

33. E.N. Frankel, W.E. Neff and T.R. Bessler, Lipids, 1979, 14, 961.

34. E.N. Frankel, Prog.Lipid Res., 1980, 19, 1.

35. D. Cobern, J.S. Hobbs, R.A. Lucas and D.J. Mackenzie, J.Chem.Soc., 1966, (C), 1897.

36. H.W.-S. Chan, J.Am.Oil Chem.Soc., 1977, 54, 100.

37. J. Terao and S. Matsushita, J.Am.Oil Chem.Soc., 1977, 54, 234.

38. E.N. Frankel, W.E. Neff, E. Selke and D. Weisleder, Lipids,

1982, <u>17</u>, 11.

39. W.E. Neff, E.N. Frankel and D. Weisleder, <u>Lipids</u>, 1982, <u>17</u>, 780.

40. E. Mihelich, <u>J.Am.Chem.Soc.</u>, 1980, <u>102</u>, 7141.

41. W.E. Neff, E.N. Frankel, C.R. Scholfield and D. Weisleder, <u>Lipids</u>, 1978, <u>13</u>, 415.

42. J. Terao and S. Matsushita, <u>Agric.Biol.Chem.</u>, 1975, <u>39</u>, 2027.

43. H.W. Gardner, <u>J.Agric.Food Chem.</u>, 1975, <u>23</u>, 129.

44. J.A. Khan and N.A. Porter, <u>Angew.Chem.Suppl.</u>, 1982, 513. K.

45. K. Miyashita, K. Fujimoto and T. Kaneda, <u>Agric.Biol.Chem.</u>, 1982, <u>46</u>, 751.

46. K. Miyashita, K. Fujimoto and T. Kaneda, <u>Agric.Biol.Chem.</u>, 1982, <u>46</u>, 2293.

47. S.S. Chang and F.A. Kummerow, <u>J.Am.Oil Chem.Soc.</u>, 1953, <u>30</u>, 251.

48. O.C. Johnson, S.S. Chang and F.A. Kummerow, <u>J.Am.Oil Chem.Soc.</u>, 1953, <u>30</u>, 317.

49. E.N. Frankel, C.D. Evans and J.C. Cowan, <u>J.Am.Oil Chem.Soc.</u>, 1960, <u>37</u>, 418.

50. C.D. Evans, E.N. Frankel, P.M. Cooney and H.A. Moser, <u>J.Am.Oil Chem.Soc.</u>, 1960, <u>37</u>, 452.

51. E.N. Frankel, C.D. Evans, H.A. Moser, D.G. McConnell and J.C. Cowan, <u>J.Am.Oil Chem.Soc.</u>, 1961, <u>38</u>, 130.

52. W.E. Neff, E.N. Frankel, E. Selke and D. Weidleder, <u>J.Am.Oil Chem.Soc.</u>, 1983, <u>60</u>, 687 (Abstract 36).

53. C.D. Evans, G.R. List, A. Dolev, D.G. McConnell and R.L. Hoffmann, <u>Lipids</u>, 1967, <u>2</u>, 432.

54. H.W.-S. Chan, F.A.A. Prescott and P.A.T. Swoboda, <u>J.Am.Oil Chem.Soc.</u>, 1976, <u>53</u>, 572.

55. E.N. Frankel, W.E. Neff and E. Selke, <u>Lipids</u>, 1981, <u>16</u>, 279.

56. M.M. Horikx, <u>J.Appl.Chem.</u>, 1965, <u>15</u>, 237.

57. P. Schieberle and W.J. Grosch, <u>J.Am.Oil Chem.Soc.</u>, 1981, <u>58</u>, 602.

58. S.S. Chang, T.H. Smouse, R.G. Krishnamurthy, B.D. Mookherjee and B.R. Reddy, <u>Chem.Ind.,Lond.</u>, 1966, 1926.

59. D.A. Forss, <u>Prog.Chem.Fats Other Lipids</u>, 1972, <u>13</u>, 177.

60. E.N. Frankel and W.E. Neff, <u>Lipids</u>, 1979, <u>14</u>, 39.

61. D.K. Park, J. Terao and S. Matsushita, <u>Agric.Biol.Chem.</u>, 1981, <u>45</u>, 2071.

62. H.T. Badings, Ned.Melk-Zuiveltijdschr., 1970, 24, 61.
63. P.A.T. Swoboda and K.E. Peers, J.Sci.Food Agric., 1978, 29, 803.
64. E. Selke, W.K. Rohwedder and H.J. Dutton, J.Am.Oil Chem.Soc., 1980, 57, 25.
65. E.N. Frankel, W.E. Neff and E. Selke, Lipids, 1983, 18, 353.
66. E.N. Frankel, "Symposium on Foods: Lipids and Their Oxidation", Avi Publishing Co., Westport, CT, USA, 1962, p.51.
67. D.A. Lillard and E.A. Day, J.Am.Oil Chem.Soc., 1964, 41, 549.
68. P.A.T. Swoboda and C.A. Lea, J.Sci.Food Agric., 1965, 16, 680.
69. W.I. Kimoto and A.M. Gaddis, J.Am.Oil Chem.Soc., 1969, 46, 403.
70. S.S. Lomanno and W.W. Nawar, J.Food Sci., 1982, 47, 744.
71. A.S. McGill, R. Hardy, J.R. Burt and F.D. Gunstone, J.Sci.Food Agric., 1974, 25, 1477.
72. G. Hoffman and P.W. Meijboom, J.Am.Oil Chem.Soc., 1968, 45, 468.
73. A.E. Wasserman, J.Agric.Food Chem., 1972, 20, 737.
74. P.W. Meijboom, J.Am.Oil Chem.Soc., 1964, 41, 326.
75. T. Yamato, T. Kurata, H. Kato and M. Fujimaki, Agric.Biol.Chem., 1970, 34, 88.
76. P. Schreier, "Flavour '81", Walter de Gruyter, New York, 1981.
77. M. Karel, "N-autoxidation in Foods and Biological Systems", Plenum Press, New York, 1980, p.191.
78. P. Brandt, E. Hollstein and C. Franzke, Lebensmittel.-ind., 1973, 20, 31.
79. N. Krog and O. Tolboe, Fette Seifen Anstrichm., 1963, 65, 732.
80. F. Linow and G. Mieth, Nahrung, 1976, 20, 19.
81. F. Tomioka and T. Kaneda, Yukagaku, 1974, 23, 777.
82. N.E. Jewell and W.W. Nawar, J.Am.Oil Chem.Soc., 1980, 57, 398.
83. A.L. Tappel, "Symposium on Foods: Lipids and Their Oxidation", Avi Publishing Co., Westport, CT, USA, 1962, p.122.
84. P.J. O'Brien and A. Rahimtula, J.Agric.Food Chem., 1975,

23, 154.

85. P.J. O'Brien, "Autoxidation in Food and Biological Systems", Plenum Press, New York, 1980, p.563.

86. L.R. Brown and K. Wuethrich, Biochim.Biophys.Acta, 1977, 464, 356.

87. M.J. Fishwick, J.Agric.Food Agric., 1970, 21, 160.

88. S. Matsushita, J.Agric.Food Chem., 1975, 23, 150.

89. W.L. Hawkins and H. Sautter, Chem.Ind.,Lond., 1962, 1825.

90. R.J. Braddock and L.R. Dugan Jr., J.Am.Oil Chem.Soc., 1973, 50, 343.

91. J. Pokorny and I. Davidek, Acta Aliment.Polon., 1979, 5, 87.

92. J. Pokorny, T.L. Nguyen, S.S. Kondratenro and G. Janicek, Nahrung, 1976, 20, 267.

93. M.W. Montgomery and E.A. Day, J.Food Sci., 1965, 30, 828.

94. I. Katz, Food Sci., 1981, 7, 217.

95. M. Maleki, Fette Seifen Anstrichm., 1973, 75, 103.

96. M. Morita, T. Aonuma and I. Nobuko, Agric.Biol.Chem.(Tokyo), 1976, 40, 2491.

97. K. Kawashima, H. Itoh and I. Chibata, J.Agric.Food Chem., 1977, 25, 202.

98. J.E. Hodge, J.Agric.Food Chem., 1953, 1, 928.

7
The Chemistry of Meat Flavour

By David A. Baines and Jerzy A. Mlotkiewicz
DALGETY SPILLERS LTD., RESEARCH AND TECHNOLOGY CENTRE, STATION ROAD,
CAMBRIDGE, U.K.

INTRODUCTION

The study of meat flavour started in the early 1950's and was almost exclusively devoted to the identification of the non-volatile, water-soluble precursors of meat flavour. Early successes were recorded and some of the key precursors, such as cysteine and ribose,[1] were identified, but it quickly became apparent that meat flavour formation was a more complex subject than had otherwise been previously anticipated and involved the complex heat interaction of many different chemical components derived from protein, lipid and free water-soluble systems such as carbohydrates. It was not surprising, therefore, that with the development of sophisticated instrumentation and separation techniques, the focus of attention in the 1960's and early 1970's turned to the identification of the aroma volatiles of meat and of the model systems formed by reacting the known precursors and their intermediate breakdown products. Towards the end of the last decade a large list of volatile compounds present in meat and model systems had been accumulated and, although a number of these compounds are known to be important contributors to meat flavour, the flavour of meat can still not be convincingly reproduced. It is now obvious that there are not one or two key character impact compounds formed during the heating process, nor that one specific precursor breaks down or reacts to form meat flavour, but that there is a complex labyrinth of reaction pathways leading to a multitude of volatiles, many of which interact with each other to form a chemical profile which we recognise, through our olfactory and gustatory senses, as meat. In addition the importance of sulphur in meat flavour chemistry has been well established, and as most sulphur chemicals exhibit very low odour thresholds, it must therefore be accepted that many of the sulphur compounds contributing to meat flavour are present in meat at such low levels that they have, to date, eluded the scrutiny of the most diligent research workers. In this decade, so far, we have witnessed a more systematic assessment of the identified meat volatiles,[2] in an attempt to pinpoint key contributors and important combinations of flavour chemicals. Over the past four years a number of excellent reviews have been written on this subject.[3-7] This logical progression of research over the past thirty years has not, however, been the only approach made in the investigation of meat flavour. The Japanese have taken a great interest in the identification of the key taste components of meat throughout this century[8] and this has led to the development of a successful industry in meat flavour enhancers in Japan following the discovery of the flavour effects of the ribonucleotides and monosodium glutamate.

This review will consider meat flavour in two parts representing the two parallel lines of investigation. The first part will be concerned with the taste components, their relationships to one another and how they influence other flavour chemicals. The second part will devote itself to the volatile flavour compounds and how they are derived from precursor systems during the cooking process. Consideration has been given only to those taste compounds and volatile flavour chemicals identified in the boiled or roasted carcass meats of beef, pork and mutton. Compounds identified in the species offals and in cured, smoked and irradiated meats have been disregarded. By taking this approach it is hoped that some tangible comparisons can be made between the species, but it must be emphasised that the data still represent different cuts of meat being heated in a variety of ways, the volatiles extracted by several methods, and separated and analysed under a variety of conditions. Nevertheless, there are still some important comparisons which can be made.

TASTE COMPONENTS OF THE RED MEATS

The water-soluble non-volatile taste components important in meats influence the basic taste sensations of saltiness, sweetness, sourness, bitterness and 'umami' in an interactive manner dependent upon environment and conditions (pH, concentration, etc.) 'Umami' is a Japanese word used to describe the gustatory effect of certain specific taste enhancers and can probably be best translated into English as savouriness, succulence and deliciousness. Kuninaka[8] defines the word as 'unique palatable tastes which do not belong to the four basic tastes'. The two classes of umami compounds defined by Boudreau et al.[9] as $umami_1$ and $umami_2$ are certain five carbon amino acids and a number of 5'-ribonucleotides. In the context of meat, the $umami_1$ effect is produced by glutamic acid and its sodium salt, and the $umami_2$ effect by the 5'-ribonucleotides, inosine-5'-monophosphate (IMP) and guanosine-5'-monophosphate (GMP).[10] The relationship between chemical structure and flavour activity of these 5'-ribonucleotides was established by Kuninaka[11] in 1960 (Fig.1). He established that the hydroxyl group on the 6-carbon of the purine ring and the phosphate ester on the 5'-carbon of the ribose moiety are both essential for flavour action.

The occurrence levels of IMP and GMP in the red meats has been measured by a number of workers[12,13] (Table 1). It is interesting to note that IMP is present in meats in relatively high concentrations and is derived from adenosine-5'-triphosphate (ATP) via adenosine-5'-monophosphate (AMP) through the action of the enzyme adenylate deaminase. Comparatively, the levels of free L-glutamic acid, the $umami_1$ enhancer, in meats is much lower than that of IMP.[14] Depending upon the part of the animal chosen, beef and pork contain between 10-35 mg/100 g of L-glutamic acid, whereas lamb is reported to be particularly rich in this amino acid and can contain double that of either beef or

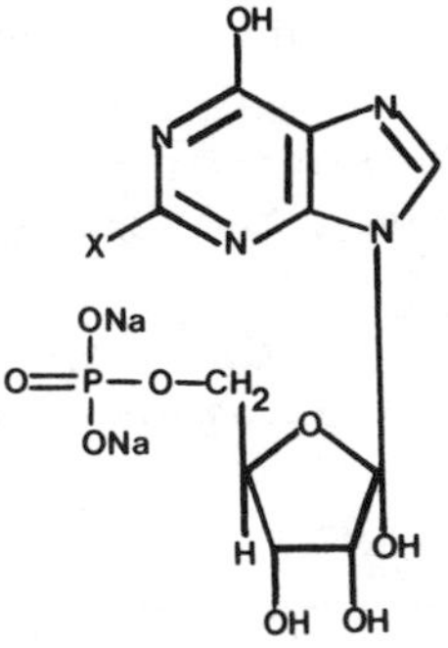

X = H; IMP   X = NH_2; GMP

Figure 1   Important 5'-ribonucleotides in meats

Table 1 Levels of IMP and GMP in the red meats (mg/100g)

	IMP	GMP	Reference
Beef	107	2.1	12
"	163	–	13
"	150	8.0	12
Pork	123	2.5	12
"	186	3.7	13
Mutton	83.5	5.1	12

pork. The threshold levels of the umami enhancers have been measured by a number of workers[15-17] and the ranges quoted indicated in Table 2.

IMP is thus present in meats at levels well above its recorded threshold range and will be playing an important role in taste appreciation, whereas GMP is present at levels on the verge of its recorded threshold range.

However, both GMP and IMP exhibit strong synergistic flavour enhancing effects with monosodium glutamate (MSG)[8] and other compounds found in meat. This synergistic effect serves to lower, sharply, the active thresholds of the 5'-ribonucleotides, as indicated in Table 3, and hence it is possible that GMP may be having a flavour taste effect at levels below its threshold.

Even more pronounced synergistic properties have been claimed in several patents for mixtures containing IMP and MSG with the individual amino acids glycine, L-serine, L-alanine, DL-tryptophan, L-methionine, L-histidine-HCl and L-aspartate[18-21] and the dipeptides asparagine-L-aspartate and ornithine-L-aspartate.[21]

Table 2 Threshold levels of 5'-ribonucleotides and MSG

Enhancer	Threshold range (mg/100 g)
IMP	10-25
GMP	3.5-20
MSG	14-30

Table 3 Synergistic effect of IMP and GMP with MSG

Enhancer	Threshold (mg/100 g)
IMP (0.1% MSG Sol^n)	0.1
GMP (0.1% MSG Sol^n)	0.03

In addition, the flavour enhancing and synergistic effects of the 5'-ribonucleotides have been found to influence an even wider range of flavour chemicals found in meat. In 1964, Kurtzman and Sjostrom[22] reported that IMP enhances desirable aroma and taste sensations such as meaty and brothy, and suppresses undesirable aroma and taste notes such as sulphury and hydrolysed vegetable protein. More recently, in 1982, Kirova and Peschevska[23] have investigated the relationship between certain indices of the sensory profile of beef broth and the levels of the 5'-ribonucleotides and have established that a significant correlation exists between the content of nucleotides, especially IMP, and 'liked' taste indices such as 'pleasant beef broth taste'. Schinneler et al.[24] report that GMP lowers the flavour threshold of octanal from 1.38 ppb in water to 0.86 ppb in water containing 10 ppm GMP. Clearly, therefore, the umami compounds are very versatile natural potentiators capable of enhancing the flavour effects of other important non-volatile taste and volatile aroma components of meats whilst imparting their, yet unknown, individual taste sensation.

A further compound which may induce a umami effect was isolated from papain tenderised beef by Yamasaki and Maekawa.[25] The compound is an octapeptide found at a level of 40 mg/100 and is described as having a delicious taste. The amino acid sequence was determined by degradation techniques followed by synthesis as:

Lys-Gly-Asp-Glu-Glu-Ser-Leu-Ala

Sensory analysis revealed that the various taste sensations elicited by the compound and their relative contributions are sweet 10%, sour 30% and savoury 60%.[26]

It is difficult to pinpoint, from the data available in the literature, which other non-volatile compounds play a role in the taste of meats as many systems have been implicated. In 1982, Motono[13] reported a study of the taste factors influencing beef by using an omission testing technique and he concluded that the key taste components are the nucleotide, IMP; the amino acids, glutamic acid, alanine, threonine and lysine; and the quaternary ammonium compound betaine. He eliminated compounds such as succinic acid, lactic acid, orthophosphoric acid, pyrrolidone carboxylic acid, anserine, carnosine, arginine and histidine considered by other workers[12,27] to make a contribution to meat taste. These compounds may, however, play a subtle role in taste differences between species and influence the general mouthfeel of meat.

The mouthfeel of meat is also strongly influenced by the fat content, and recently Korova and Peschevska[23] established that a significant relationship exists between total lipids and the parameter 'fatty taste' and between free fatty acids and 'pleasant taste of beef broth'. An inverse relationship was found between the lipid content and 'pleasant aroma of cooked beef' but no correlation at all was found between total carbonyls and any of the measured sensory parameters.

In conclusion, the taste of meat appears to be influenced by the nucleotide content, certain amino acids, peptides, lipids and free fatty acids. Variations in the levels of these compounds will have an influence on species character but the greatest variation is observed in the ratios of the free fatty acids and this is described in detail in the following section.

VOLATILE AROMA COMPONENTS OF THE RED MEATS

Over the past two decades a bewildering array of organic volatiles has been identified in the red meats. The total number, subdivided into their various chemical classes for beef, pork and lamb, are listed in Table 4. Due to the imbalance created by the large number identified in beef, the volatiles in each class are also recorded as a percentage of the total volatiles to allow species comparisons to be made. The impressive total of 662 compounds found in beef reflects the interest shown by researchers in this meat in comparison to other species, and mirrors the consumer preference for beef. Hsu et al.[28] alone have added a further 67 compounds to the beef list in 1982. These figures must, however, be viewed with some caution because many of the compounds identified in beef have been detected simply because more work has been done with this meat, and the differences observed may not be genuine species differences. Additionally, very few workers have quoted the levels of compounds detected and thus a compound identified in one meat may be present at a level which is well below its threshold and play little or no role in the flavour profile, but in another meat it may be present at a level which gives it

Table 4 Volatile compounds found in meat

Class of compound	Beef number	Beef %total	Pork number	Pork %total	Lamb number	Lamb %total
Aldehydes	56	8	33	14	43	18
Ketones	56	8	14	6	23	10
Acids	26	4	40	17	47	20
Alcohols	55	8	25	11	14	6
S-Non cyclic	73	11	26	11	6	3
Thiophenes	39	6	3	1	3	<1
Furans	37	6	8	3	6	3
S-cyclic	15	2	4	2	5	2
Esters	30	5	29	12	5	2
Hydrocarbons	120	18	30	13	33	14
Pyridines	10	2	2	1	16	7
Pyrazines	48	7	5	2	14	6
Pyrimidines	2	<1	-	-	-	-
Lactones	32	5	12	5	14	6
Thiazoles	16	2	1	<1	4	2
Oxazoles	9	1	-	-	-	-
Pyrroles	12	2	-	-	1	<1
Amines	3	<1	1	<1	2	<1
Ethers	5	1	-	-	-	-
Chloro compounds	8	1	2	1	-	-
Miscellaneous	10	2	2	1	-	-
Total	662	∿100	237	∿100	236	100

The data on this table were acquired from the following references: Beef 3, 28-32; Pork 31, 33-37; Lamb 38, 39

significance.   Variability in the extraction techniques employed
may also distort the figures.   One process may favour, or be
designed to favour, the isolation of a particular class of
chemicals whereas another process may miss them completely.
However, by publishing the data in this form, some of the more
obvious species differences can be highlighted and these will be
reviewed in the following sections.   One thing is very clear from
Table 4; that given this wealth of data it is apparent that no
one single compound is uniquely responsible for the aroma of
cooked meat. Several do, however, have an important role to play
in the various meat species and these, and their formation
pathways in the cooking process, are reviewed.   It is beyond the
scope of this review to list all the compounds identified in
meat; further information about individual compounds and groups
of compounds can be obtained from the referenced authors and in
particular MacLeod and Seyyedain-Ardebilli[3] who have documented
the majority of compounds detected in beef.

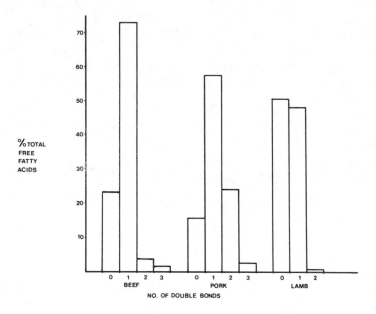

Figure 2 Distribution of free fatty acids in beef, pork and lamb
after heating in air at 100°C

Fatty acids, esters, lactones

Hornstein and Crowe[40],[41] were the first to investigate the distribution of free fatty acids in beef, pork and lamb and they demonstrated that significant differences exist in the proportion of the unsaturated free fatty acids between these meats after heating in air at 100°C (Fig.2). Lamb meat was found to contain only a small proportion of free fatty acids with two or more double bonds but contains a high percentage of saturated fat which possibly contributes to its unpopularity as a cold meat. Beef contains approximately 5% of free fatty acids with two or more double bonds, but pork contains over 26% of these fatty acids. Since this early work, the total number of free fatty acids identified in the three red meat species has risen considerably (Table 5).

Not only is it obvious from Table 5 that the free fatty acids are a high proportion of the total volatiles detected in pork and lamb, but that in lamb the branched chain saturated fatty acids predominate. A comprehensive study of these branched acids in lamb meat/fat has been made by Wong et al.[42] who attribute the low consumer acceptance of sheep meat in many countries to 4-methyloctanoic acid and, to a lesser extent, 4-methylnonanoic acid. Asians, for example, find the flavour of mutton distasteful, and the Chinese have a word 'soo' which is used to describe specifically the characteristic flavour of cooked mutton. 4-Methyloctanoic acid at a concentration of

Table 5 Free fatty acids in the red meats

	Beef	Pork	Lamb
Straight chain saturated	15	15	10
Branched chain saturated	3	4	19
Unsaturated	7	13	14
Dicarboxylic	–	4	–
Aromatic	–	1	2
Straight chain oxygenated	1	3	2
Total	26	40	47
% of total volatiles	4	17	20

0.5 ppm was found to create a positive association of muttoniness
and 'soo' when administered with mince. Lamb also contains
methyl and dimethyl substituted acids on the 2,4,6,8,9 and 10
carbons and also an enhanced content of odd numbered n-fatty
acids. Table 5 also highlights the large differences that exist
in the distribution of unsaturated fatty acids in the red meats.
Pork and lamb contain a similar number, but only half that number
have been found in beef. The differences in pork and lamb are
accounted for by a number of short chain unsaturated fatty acids
which do not appear to be present in beef. Their contribution to
the flavour of meat and meat fat will be small but they may
contribute fatty, metallic and green flavour notes.

The free fatty acids in meat are derived from triglycerides
and phospholipids by either the action of bacterial enzymes or by
hydrolysis and thermal oxidation during the cooking process. The
branched chain acids in lamb are known to occur via fatty acid
synthesis utilising methyl malonate and are accumulated through
the inefficient metabolism of propionate derived from the
diet.[43] The animal feeding regime also influences the relative
distribution of the free fatty acids within the lipid fraction
and hence can influence consumer appreciation. Van den
Ouweland[44] points out that the free fatty acids reside mainly in
the intramuscular lipid fraction (tissue lipids), which produces
the marbling effect in meats, and that it is this lipid fraction
which accounts primarily for the species differences. This is
borne out by Hornstein and Crowe[40] and other workers[45] who have
demonstrated that, although the flavour generated from the lean
is meaty, the same meaty flavour is produced from the lean meat
of different species.

Wasserman and Spinelli[46] undertook an organoleptic study of
the lipids of beef, pork and lamb and concluded that the lipids
of pork and lamb were recognisable as their species even when
these lipids were heated over prolonged periods, but the beef
lipid fraction resembled hot fat and was not readily
identifiable. The intermuscular lipids (depot fat) play an
important role in flavour retention. They act as a solvent for
the aroma volatiles formed in the lean and serve not only as a
reservoir for these flavour chemicals but as a site for further
reaction.[44] It is here, for instance, that esters are formed
from the interaction of free fatty acids and alcohols generated

by lipid oxidation. The distribution of esters between the species is outlined in Table 6 and shows that pork contains a high proportion. Kevei and Kozma[47] have reported that raw pork muscle contains only a small number of esters but cooked pork contains significantly more, with acetates being the most

Table 6 Esters of the red meats

	Beef	Pork	Lamb
Aromatic	3	–	–
Aliphatic acetates	7	10	–
" others	16	19	5
Miscellaneous	4	–	–
Total	30	29	5
% of total volatiles	5	12	2

numerous. Eighteen of the compounds identified in cooked pork are derived from C1-C10 acids which tend to impart a fruity sweet note to pork meat. Beef, on the other hand, contains a higher proportion of esters derived from long chain fatty acids which possess a more fatty flavour character. Interestingly, no esters of unsaturated acids have been found in the cooked meats of the three species although they have been identified in cured meats.

The lactones identified in the different meats are classified in Table 7. They are formed in the lipid portion by the lactonisation of γ- and δ-hydroxy fatty acids, which are normal constituents of triglycerides, and from the oxidation of oleic acid and unsaturated aldehydes. The lactones listed under beef were largely isolated from beef fat and include the range of odd- and even-numbered saturated γ- and δ-lactones ranging from C4 to C16.

The predominant lactones in all three red meats are the δ-long chained compounds, δ-12, 14 and 16 with δ-tetradecalactone predominating. The flavour properties of these lactones are buttery, oily, fatty and fruity and they contribute in a minor way only to the overall character of meat. The sulphur substituted lactone found in beef[48] is β-methylthio-γ-

Table 7 Lactones identified in red meats

	Beef	Pork	Lamb
γ-lactones	15	8	8
δ-lactones	13	4	5
Unsaturated lactones	3	–	1
Sulphur substituted lactones	1	–	–
Total	32	12	14
% of total volatiles	5	5	6

butyrolactone (1) formed by the Michael addition of methanethiol to 2-butenolide (2) which has also been detected in beef (Fig.3).

When lambs are fed a dietary supplement of lipid-protected sunflower seed, the meat shows significantly enhanced levels of linoleic acid and possesses a very different but objectionable flavour described as sweet and oily. The compound responsible for imparting this flavour has been identified by Park et al.[49] as 4-hydroxy-dodec-cis-6-enoic-γ-lactone. It was found along with its saturated counterpart and a higher than normal level of trans,trans-2,4-decadienal, a predominant oxidation product of linoleic acid.

Aldehydes and ketones

The oxidative decomposition of lipid-derived fatty acids during the cooking process leads to the formation of significant quantities of the full range of straight chain alkanals and a wide range of alkenals. The fatty acids most susceptible to this autoxidative process are those with active methylenes adjacent to

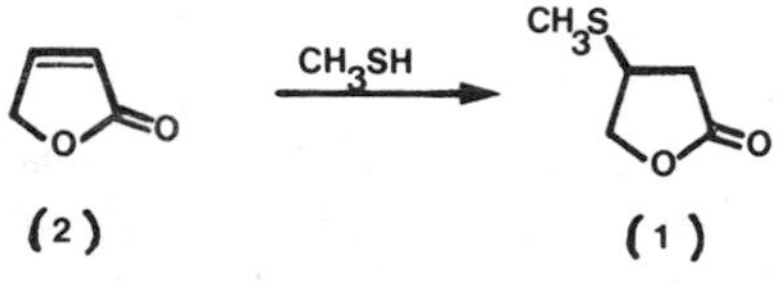

(2)   **(1)**

Figure 3 The formation of β-methylthio-γ-butyrolactone (1).

double bonds and hence in meat systems oleic, linoleic, linolenic and arachidonic acids are the most prone to oxidative attack and aldehyde formation. The process is catalysed in meat by traces of transition metals and by haemo compounds and, in addition to aldehydes, ketones, alcohols and hydrocarbons are formed. The reactions are complex and, for example, it is postulated that linoleic acid (3) can yield seven individual mono-hydroperoxide intermediates on carbon atoms 8 to 14.[50] Fig.4 shows how these hydroperoxide intermediates can be formed from the three resonating systems (4), (5) and (6) and gives as example the formation of hexanal (7) produced in the highest quantity in this

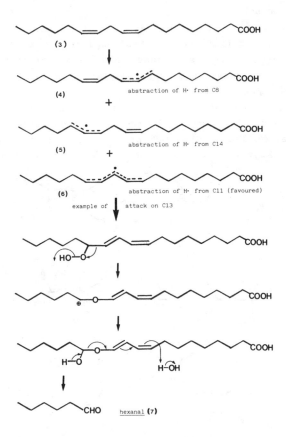

Figure 4 Autoxidation of linoleic acid

reaction. Badings[50] has made a study of the volatile carbonyl
compounds resulting from the autoxidation of oleic, linoleic,
linolenic and arachidonic acids and has found a large number of
aldehydes and ketones, most of which have been identified in
meat.

Loury[51] has demonstrated that during autoxidation even the
saturated aldehydes such as decanal break down stepwise to form a
complete set of lower homologues. In meat fat this process
probably accounts for the presence of those aldehydes which
cannot be formed by a breakdown of the hydroperoxides derived
from the unsaturated fatty acids. Fuller details of these
processes are provided in the recent excellent review on the
subject by Grosch.[52]

The straight chain aldehydes are found in both raw and
cooked meats, but branched chain aldehydes have been found only
in cooked meats. This is because their major formation route is
by the oxidative deamination-decarboxylation of α-amino acids via
the Strecker degradation which takes place during the cooking
process. The aldehydes produced by this route correspond to the
carbon chain of the amino acid minus the carbon atom lost in the
decarboxylation step. The mechanism of the Strecker degradation
is shown in Fig.5. A list of the volatile aldehydes produced in

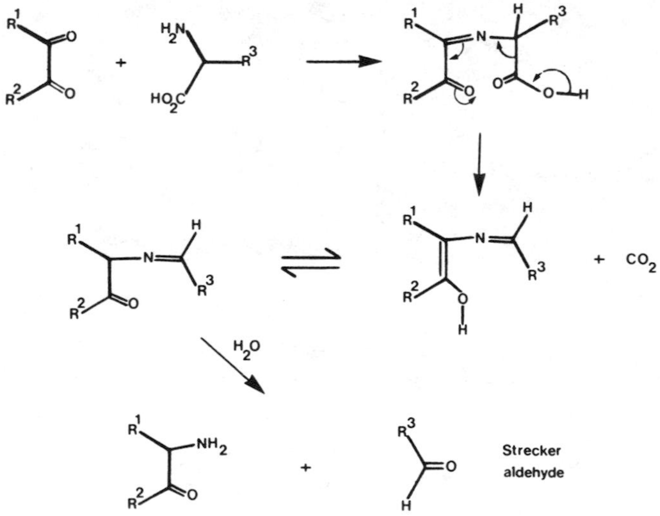

Figure 5 Mechanism of the Strecker degradation

Table 8 Aldehydes produced in the Strecker degradation

Original amino acid	Aldehyde formed	Presence of aldehyde in meat species		
Glycine	formaldehyde	beef	pork	–
Alanine	acetaldehyde	beef	pork	lamb
Cysteine	acetaldehyde	beef	pork	lamb
"	mercaptoacetaldehyde	–	–	–
2 aminobutyric acid	propanal	beef	pork	lamb
Norvaline	butanal	beef	pork	lamb
Valine	2-methylpropanal	beef	–	lamb
Norleucine	pentanal	beef	pork	lamb
Leucine	3-methylbutanal	beef	pork	lamb
Iso leucine	2-methylbutanal	beef	pork	lamb
Phenylalanine	phenylethanal	beef	pork	lamb
Glutamic acid	3-oxobutyric acid	–	–	–
Methionine	methional	beef	–	–

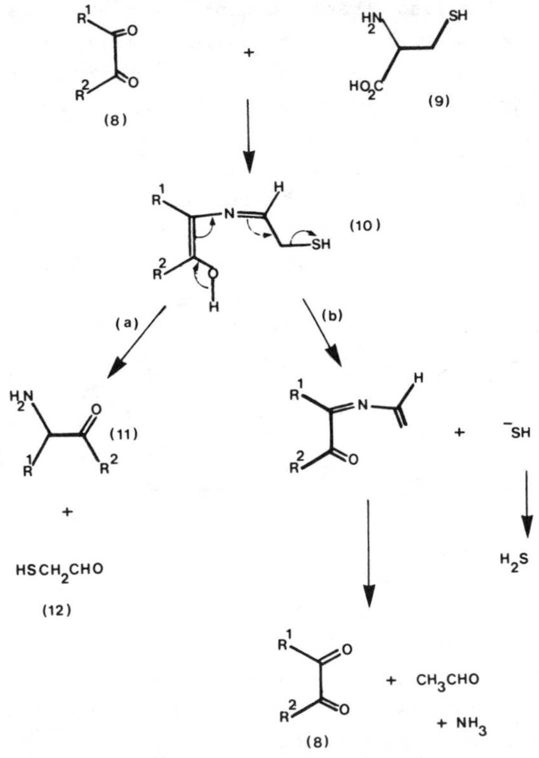

Figure 6 Strecker degradation of cysteine

the Strecker degradation and their detection in the species of meats is shown in Table 8. Stressing the importance of the sulphur amino acids in meat flavour, in the Strecker degradation of cysteine (9), the intermediate imine-enol (10) can break down by two pathways (Fig.6). The normal pathway (route a) leads to the formation of mercaptoacetaldehyde (12), a useful reactive intermediate which leads to the formation of thiazolines and thiazoles and the amino ketone (11) which, by self-condensation, forms pyrazines. The second pathway (route b) produces hydrogen sulphide, ammonia and acetaldehyde plus regeneration of the original diketone (8). Thus cysteine is a particularly important source of a number of reactive intermediates, especially hydrogen sulphide, which go on to play an essential role in the formation of meat flavour chemicals.

Nearly all of these compounds have been found in red meats. One would expect also that, judging by the large number of branched chain fatty acids in lamb, similar branched chain aldehydes would be detected unique to lamb, but none has been found. The various sub-classes of aldehydes identified in the red meats are indicated in Table 9.

Table 9 Aldehyde distribution in the red meats

	Beef	Pork	Lamb
Straight chain	18	16	14
Branched chain	6	3	3
Unsaturated	22	9	24
Aromatic	7	4	2
Oxygenated	3	1	1
Total	56	33	43
% of total volatiles	8	14	18

From Table 9 it is unclear why more unsaturated aldehydes have not been found in pork, considering the preponderance of unsaturated fatty acids in this meat. This probably reflects the relatively greater interest that has been taken in cured pork meats as opposed to the uncured meat, where more unsaturated aldehydes have been detected. In terms of concentration, Hornstein and Crowe[40,41] have measured the amount of total carbonyls isolated from beef, pork and lamb and have determined that they are present in the ratio 5:14:1 respectively. Mottram[31] quotes that the volatiles from beef and pork are dominated by aliphatic aldehydes which contribute between 30-50% of the total in terms of the gas chromatographic peak areas of the headspace. He has also shown that there is ten times more hexanal in pork (plus pork fat) than in beef (plus beef fat).

The contribution of aldehydes to the aroma of meat is difficult to assess and the subject of much debate. Wong et al.[42] state that aldehydes in lamb play a minor flavour role, but Caporaso[53] implicates ten aldehydes as being flavour significant in this meat. Trans-2-trans-6-nonadienal has been identified to be responsible for a tallowy flavour in beef and mutton fat[39] and has been found in all three red meats. Likewise, trans-2-cis-4 and trans-2-trans-4-decadienals, the dominant dienals in the red meats, possess a deep fat fried aroma and have low thresholds in water (0.02 and 0.1 ppm respectively). Kirova and Peschevska[23] have recently attempted to measure how a sensory panel perceives the aroma of volatile carbonyl compounds in beef and attribute their contribution as adding rancid, caramel, fatty and mushroom aroma notes. A recent patent filed by the Firmenich[54] company records the importance of a number of longer chain branched unsaturated aldehydes to the flavour of beef. One such compound, 6-methylhept-2-enal, has been found in beef by Yamato et al.[55] and by Chang and Peterson[56] but this was not implicated in the patent. The most important role that aldehydes play in meat flavour is that of building blocks for many more important flavour contributors such as thiazoles and cyclic sulphur compounds. Formed during the cooking process, the electrophilic aldehydic carbon is competed for by the abundance of nucleophilic scavengers (H_2S, NH_3, CH_3SH, etc.) generated in the lean meat by decomposition of amino acids.

Turning briefly to ketones, of the number identified in the red meats the most abundant are the alkan-2-ones formed by the pyrolysis of triglycerides containing a β-keto fatty acid. The most important ketones are the α-dicarbonyls such as pyruvaldehyde and diacetyl formed in the Maillard reaction. These act as building blocks for pyrazines, thiazoles, oxazoles, etc., as well as being involved in Strecker degradations. One particular ketone, the steroid 5α-androst-16-en-3-one is responsible for the 'boar taint' aroma associated with pork fat.[57] Also implicated as contributing to this taint are skatole and indole.[34]

Furans and thiophenes

The furans and thiophenes are probably the most important but most elusive members of the family of meat flavour chemicals. A great number have been identified in beef but comparatively few in both pork and lamb (Table 10). The furans are derived in meat from several sources, the most important of which is the Amadori rearrangement involving the interaction of carbohydrates with amino acids, a reaction which is part of the overall Maillard process. Other defined routes to furans include the oxidation of unsaturated aldehydes, the thermal decomposition of thiamine and the breakdown of 5'-ribonucleotides (Fig.7). In model reactions, Shibamoto and Russell[58],[59] have demonstrated how both furans and thiophenes can be generated from the interaction of glucose with ammonia and hydrogen sulphide. Thiophenes are also formed from the thermal decomposition of thiamine, the reaction of sulphur-containing amino acids with carbohydrates and the reaction of furans with hydrogen sulphide (Fig.7).

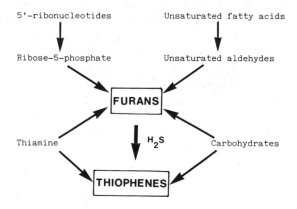

Figure 7   Generation of furans and thiophenes

The levels of these precursors of furans and thiophenes in meats have been measured. Nucleotide levels and the levels of aldehydes have previously been discussed in this Chapter. The major carbohydrates in the red meats are glucose, fructose and ribose. Beef contains around 48 mg/100 g tissue of total carbohydrates, pork 46 mg/100 g and lamb 36 mg/100 g.[60] Over 95% of the total carbohydrate fraction of the three red meats is made up of glucose, with fructose and ribose comprising the remainder. Notable differences exist in the levels of thiamine between the red meats. Lean beef contains 0.07 mg/100 g, lean lamb 0.14 mg/100 g and lean pork 0.89 mg/100 g.[61] Thus, pork contains over 10 times more thiamine than beef and 6 times more than lamb. The classes of furans and thiophenes identified in the red meats are outlined on Tables 10 and 11.

Table 10 Furans in the red meats

	Beef	Pork	Lamb
Furan	1	1	1
Alkyl substituted	15	3	2
Oxygenated side chain	12	3	3
Sulphur containing	2	1	–
Furanones	4	–	–
Bicyclic systems	2	–	–
Others	1	–	–
Total	37	8	6
% of total volatiles	6	3	3

The largest group of furans detected in beef are alkyl substituted, usually in the 2-position. These are derived from oxidised aldehydes, and pentyl furan is usually found in the highest concentration. Chang et al.[62] have suggested that this compound is obtained from the cyclisation of 4-ketononal derived from linoleic acid.

The sensory properties of the majority of furans detected in the meats do not appear to contribute in a positive manner to the flavour. They tend to be ethereal, caramellic, sickly, sweet.

Table 11 Thiophenes in the red meats

	Beef	Pork	Lamb
Thiophene	1	1	1
Tetrahydro derivatives	2	–	1
Monoalkyl	14	1	–
Dialkyl	2	–	–
Oxygenated side chain	17	1	1
Thiophenones	3	–	–
Total	39	3	3
% of total volatiles	6	1	1

The thiophenes identified tend to be oniony, burnt, sulphurous, rubbery, solventy, and likewise play a minor role although they do have some influence on the roast character of meat. Those furans and thiophenes that are felt to contribute significantly to the flavour of meat are formed in the reaction of alkylhydroxy

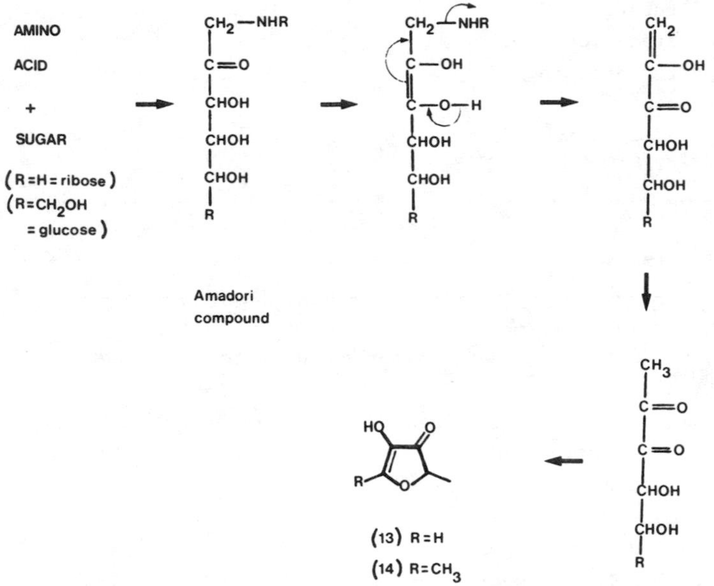

AMINO
ACID

+

SUGAR

(R = H = ribose)
(R = CH$_2$OH
= glucose)

CH$_2$—NHR
C=O
CHOH
CHOH
CHOH
R

Amadori
compound

CH$_2$—NHR
C—OH
C—O—H
CHOH
CHOH
R

CH$_2$
C—OH
C=O
CHOH
CHOH
R

CH$_3$
C=O
C=O
CHOH
CHOH
R

(13) R = H
(14) R = CH$_3$

Figure 8 Formation of furanones

furanones with either hydrogen sulphide or cysteine, but are so elusive that they have not yet been detected in meat systems. The two furanones implicated in this reaction are 4-hydroxy-5 methyl-3(2H)-furanone (13) and 4-hydroxy-2,5-dimethyl-3(2H)-furanone (14) and both have been found in beef by Tonsbeek et al.[63] The methyl homologue is derived in meat either from ribose-5-phosphate formed by the breakdown of 5'-ribonucleotides, or from free pentoses (mainly ribose) via the Amadori rearrangement (Fig.8).

The dimethyl homologue is formed from the Amadori rearrangement of glucose, the most abundant carbohydrate in meat. The reaction of 4-hydroxy-5-methyl-3(2H)-furanone (13) with hydrogen sulphide has been studied by Ouweland and Peer.[64] The first stage of the reaction is the formation of an equilibrium mixture of the furanone with its thio analogue (Fig.9). Both compounds then proceed to react further with H_2S to form a complex mixture which includes thiophenes and furans, a number of which are 3-mercapto substituted (Fig.10). The 2,5-dimethyl homologue is reported to react in an analogous manner, forming an equally interesting range of meaty compounds.

Odour descriptions of some of the reaction products are as follows:

a) Furans

    3-Mercapto-2-methyl-4,5-dihydrofuran   Roasted meat

    4-Mercapto-2-methylfuran   Green meaty

b) Thiophenes

    3-Mercapto-2-methylthiophene   Roasted meat

    3-Mercapto-2-methyl-2,3-dihydrothiophene   Sweet roasted meat

    3-Mercapto-2-methyl-4,5-dihydrothiophene   Roasted meat

    4-Mercapto-2-methyl-4,5-dihydrothiophene   Roasted meat

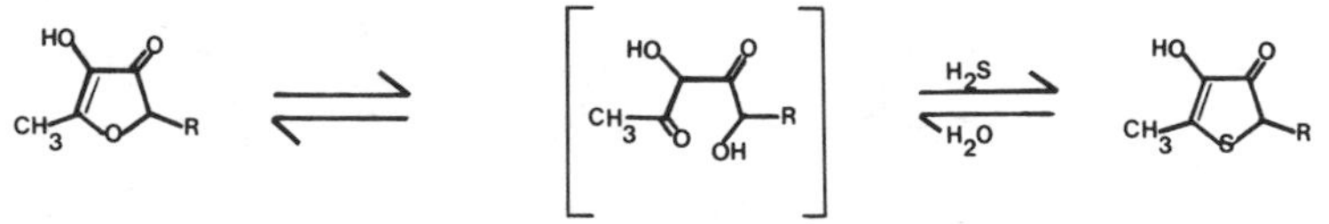

(13) R = H
(14) R = CH₃

Figure 9 Initial reaction of furanones with hydrogen sulphide

2-Mercapto substituted furans and thiophenes do not possess meaty
aroma characteristics but are burnt and sulphurous, and thus it
is the 3- and 4- substituted systems that possess the important
meaty qualities.

Evers et al.[65] made a synthetic study of 3-mercapto
substituted furans and reported equally interesting organoleptic
properties for these and related compounds. Bodrero et al.[2]
recently made a study of the contribution of flavour volatiles to
the aroma of beef by a surface response methodology technique and
found that the reaction mixture of 4-hydroxy-5-methyl-3(2H)-
furanone contributed significantly to the flavour of beef.
3-Mercapto substituted furan systems can also be formed in meat
from the thermal degradation of thiamine in a reaction which has
been studied in detail by Dwivedi and Arnold[66] and Van der Linde
et al.[67] The formation pathway to these compounds and to other
products formed in the same reaction is outlined in Fig.11.

In mildly basic conditions, the thiazole ring opens to
produce 5-hydroxy-3-mercaptopentan-2-one (15) which cyclises to
form 3-mercaptofurans (16 and 17).   Both the intermediate
mercaptopentanone and the two mercaptofurans have been identified
in the product mixture along with other furans, thiophenes,
thiazoles, 3-mercaptopentan-2-one and 3-mercaptopropanol.   In
mildly acidic conditions, the primary reaction route is the
cleavage of the methylene bridge to form 4-methyl-5-(2-hydroxy-
ethyl) thiazole (18), a compound recently detected in meat by

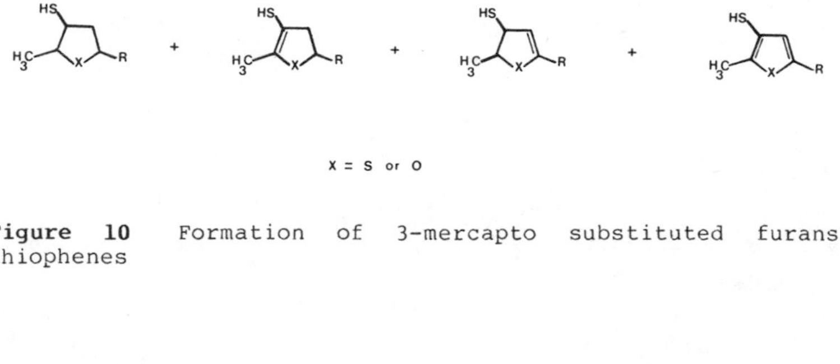

X = S or O

Figure 10   Formation of 3-mercapto substituted furans and
thiophenes

Hsu et al.[28].

Several of the reaction products resulting from the thermal degradation of thiamine have been found in meat, but so far the 3-mercapto substituted furans have not been reported. It therefore remains a matter of some speculation about their influence upon the flavour of meat, but it also indicates that compounds are present at very low concentrations which are very difficult to detect but are playing an important role. An example of this is a product resulting from the photolysis of thiamine, 1-methylbicyclo[3.3.0]-2,4-dithia-8-oxaoctane[68] (19), which is currently the subject of some controversy.[69] The

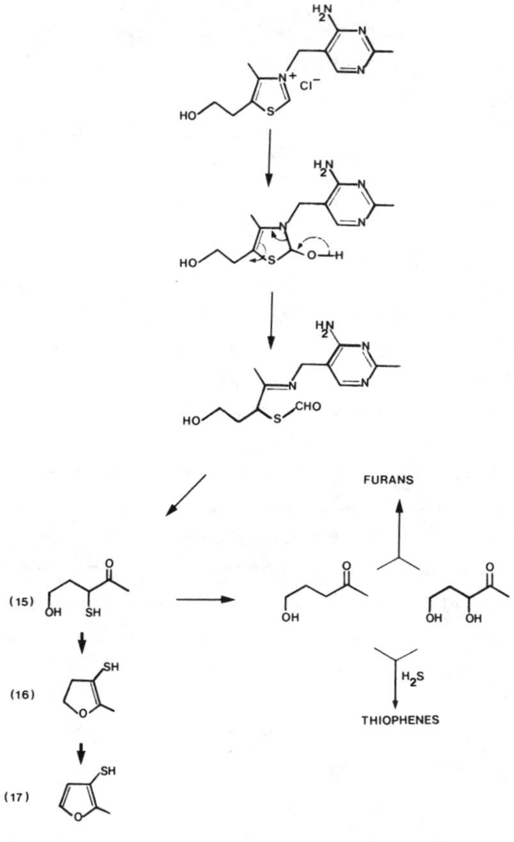

Figure 11 Thiamine degradation

(18)   (19)

threshold of the product isolated from the photolysis mixture was
measured at 4 parts in 10^{13} and is one of the lowest thresholds
ever recorded.[70] However, the pure synthetic material exhibits a
threshold of 6 parts in 10^9 and it has been concluded that the
low threshold of the product extracted from the photolysis
mixture may be due to an exceptionally potent impurity.[69]
Clearly, compounds of that organoleptic strength will be very
difficult to detect in a natural product using mass spectrometry
and, if such compounds are formed in meat, it will be necessary
to isolate the flavour volatiles from very large quantities of
the raw material and subject the extracts to arduous
fractionation.

Pyrazines

The occurrence of this class of compound in foods was not
widely reported until the mid-1960's but since then considerable
attention has been turned to the isolation and study of these key
compounds in food systems. Several groups of workers have
written comprehensive reviews on the isolation, identification,
formation and synthesis of pyrazines.[71,72]

Looking at Table 12 of pyrazine classes within each of the
three red meat species, it can easily be seen that a high number
have been identified in beef whereas pork and lamb contain a
relatively small number, lamb containing slightly more than
pork. On a percentage basis, lamb and beef are almost identical
but beef stands out in terms of actual numbers of pyrazines.

Considering alkyl substituted pyrazines (mono- to tetra-),
there are some species differences in numbers and types reported,
but the more obvious differences between the species as regards
pyrazines detected occur in the remaining classes. For example,

Table 12 Pyrazines in the red meats

	Beef	Pork	Lamb
Pyrazine	1	-	-
Monoalkyl	2	1	1
Dialkyl	9	3	5
Tri- and tetraalkyl	12	1	8
Bicyclo	16	-	-
Unsaturated sidechain	4	-	-
Acyl	4	-	-
Total	48	5	14
% of total volatiles	7	2	6

beef is unique in the fact that it contains bicyclo pyrazines, 2-acetyl pyrazines and pyrazines substituted with unsaturated side-chains. It has been reported in the literature that bicyclic pyrazines, in terms of organoleptic properties, have roasted or grilled animal notes and so could be a major contributor to the overall roast meat notes of roasted beef.[7] Flament et al.[48] have identified a series of pyrrolopyrazines of the type shown by general formula (20) where R^1, R^2 and R^3 are either H or methyl. A second interesting structural cyclic class found in beef are the dihydrocyclopentapyrazines (21) identified by Mussinan et al.[73] and Flament et al.[74] where R^1, R^2 and R^3 are either hydrogen, methyl or ethyl.

In general terms, as the degree of substitution increases on the pyrazine ring, so the organoleptic properties of the pyrazines become more nutty and roasted. Acetyl pyrazines usually have an intense and characteristic roasted note reminiscent of popcorn[75] and this must also play a role in the overall roast-type notes of roast beef. Recently, Bodrero et al.[2] evaluated a number of compounds found in cooked beef for their contribution to the aroma of beef using a surface response method and obtained the highest score from 2-acetyl-3-methyl pyrazine (22). It was described as having a slightly meaty aroma. However, the combination of 2-acetyl-3-methyl pyrazine

(20)   (21)

(22)

with hydrogen sulphide gave the highest observed score for a
mixture and was described as an exceptionally good, full meaty
aroma.   2,3-Dimethyl pyrazine also scored well in this study.
Formation pathways for pyrazines in the cooking process generally
follow either amino acid-sugar interactions or pyrolysis of
α-amino ketones produced during the Strecker degradation of amino
acids (Fig.12).   The bicyclic cyclopentapyrazines (21) are formed
from the condensation of hydroxycyclopentenones, products of the
Maillard   reaction,   with   α-dicarbonyls   in   the   presence   of
ammonia.   The pyrrolopyrazines (20) are probably the result of a
similar condensation involving the interaction of an α-dicarbonyl
compound with an α-amino carbonyl Strecker product.   There has
been considerable dispute as to the source of nitrogen in the
pyrazine ring.   Several groups report that free ammonia is
necessary for pyrazine formation.[76,77]   However, Koehler et al.[78]
demonstrated   that   different   pyrazine   distributions   were
influenced by the choice of amino acid, showing that the reactive
ammonia unit remained attached to the amino acid.   Furthermore
they showed through C[14] labelling that the sugar molecule was the
primary carbon source for pyrazine formation.
    Although some pyrazines form over long periods at -5°C, it

is generally felt that significant pyrazine formation does not begin until 70°C and increases with temperature.[5] The roasted type notes of the more heavily substituted pyrazines are associated with more severe heating and the exterior browned surface of roast meats.

In a model study of the reaction of glucose, ammonia and hydrogen sulphide 23 products were obtained, 10 of which were pyrazines, and it was found that an increasing concentration of ammonia increased the total yield of pyrazines and changed the ratio of individual pyrazines formed.[5] Thus the concentration of ammonia and total amino content of the species meats may have some bearing on the ease of formation of the pyrazines in these species. Macy et al.[60] report quantities of total amino compounds in unheated lyophilised diffusates of beef, pork and lamb as 161.53 mg/100 g, 109.86 mg/100 g and 130.01 mg/100 g respectively. Thus a tenuous relationship exists between the total number of pyrazines found in each species of meat and the total content of amino compounds. The greater interest shown in beef is another cause for the detection of a larger number of pyrazines, but also much of the accumulated data for beef is from the roasted meat, which is a cooking process favouring pyrazine formation.

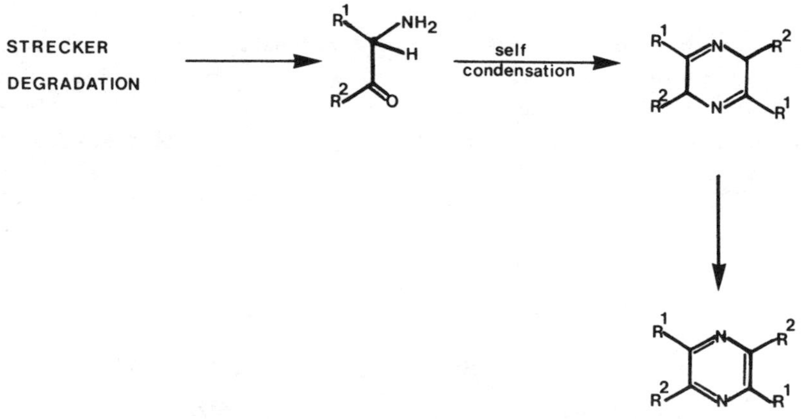

Figure 12 General pathway for the formation of pyrazines

Table 13 Pyridines in the red meats

	Beef	Pork	Lamb
Pyridine	1	1	1
Alkyl substituted	5	1	14
Oxygenated side chain	2	-	1
Nitrile substituted	2	-	-
Total	10	2	16
% of total volatiles	2	1	7

Pyridines

Pyridine formation in meat is very much a species related phenomenon (Table 13) with a large number of compounds having been detected in lamb but relatively few in beef and pork. The total of 14 alkyl pyridines identified in lamb by Buttery et al.[79] range from methyl to hexyl substituted systems through to a number of dialkyl substituted pyridines. These pyridines were identified in the steam volatile oil obtained from roasted lamb fat and Buttery et al.[79] suggest that they are derived from the condensation of dienals with ammonia. Thus pentylpyridine (24) is formed from the interaction of ammonia with 2,4-decadienal (23) derived from the oxidative breakdown of linoleic and arachidonic acids (Fig.13). It is reasonable that this should occur because more ammonia is generated in this species. It contains a higher concentration of amino acids, but a lower content of sugars, than the other species and therefore amino

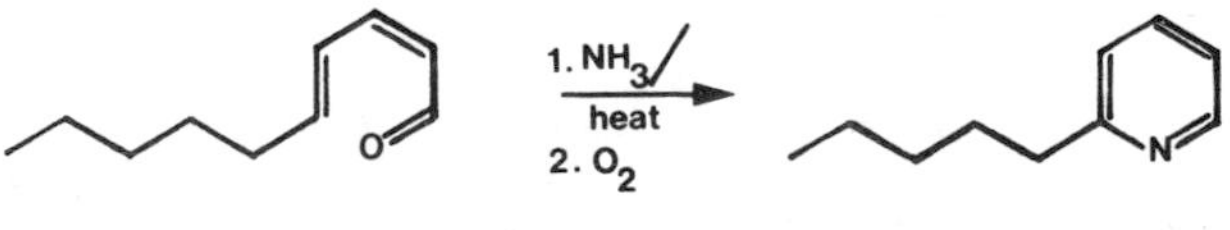

(23)   (24)

Figure 13 Formation of pentylpyridine

acids are more likely to be thermally degraded producing ammonia, than involved in the Maillard reaction where the amino nitrogen is incorporated into other systems such as pyrazines.

Other modes of formation of the pyridines in heated food systems include the thermal decomposition of alanine and β-alanine,[80] the condensation of amino acids with simple aldehydes to form quaternary pyridinium betaines which thermally decompose to pyridines and from the interaction of proline and glucose.[81] The odours of the alkylpyridines are not pleasant and are likely to detract from the flavour of lamb. Pentylpyridine is reported to exhibit a threshold of 0.6 parts per 10^9 parts of water and possesses a fatty, tallow aroma. Buttery et al.[79] suggest that the alkylpyridines of lamb fat contribute to the low popularity of this meat by many consumers.

Thiazoles and oxazoles

These two groups have been placed together because there are some similarities in their mode of formation; thiazolines and oxazolines are encompassed under the group heading. Relatively few oxazoles and oxazolines have been found in meat/food flavours and, of the three species studied in this review, beef is the only one in which these compounds have been identified. The breakdown of the constituent members of the group are three oxazoles (2,4,5-trimethyloxazole, 2-methylbenzoxazole, 4,5-dimethyloxazole), five alkyl substituted 3-oxazolines and an oxazilidindione. 2,4,5-Trimethyloxazole has a green, nutty aroma [82] and the corresponding 3-oxazoline a woody, musty, green note. In general the oxazolines have a nutty, green, woody note and make a minor contribution to the overall flavour properties of cooked beef aroma.[82]

There are several pathways proposed for the formation of oxazoles and oxazolines. One route is the decarboxylation of hydroxylated amino acids such as serine or threonine leading to β-amino alcohols which condense with aldehydes then cyclise to give oxazolidines. Oxidation of the latter compounds results in the formation of oxazoles. A second pathway is the interaction of α-amino ketones, formed during the Strecker degradation of α-amino acids, with aldehydes formed either from the Strecker degradation of amino acids or the thermal/oxidative degradation of fatty acids. The resultant condensation/cyclisation produces

oxazolines which oxidise to oxazoles. This route is very similar
to one method of formation of thiazolines and thiazoles. A third
method of formation is the thermal interaction of ammonia,
acetaldehyde and acetoin. Various other systems are proposed
which will produce oxazoles.[83]

Thiazoles and thiazolines are closely related to oxazoles
and oxazolines in structure but are a very important class of
compounds in terms of flavour and aroma. All three species have
thiazoles or thiazolines identified in them, but again beef is
prominent because of the relatively higher number of these
compounds reported. Table 14 shows the breakdown of this group.

Table 14 Thiazoles and thiazolines in the red meats

	Beef	Pork	Lamb
Thiazole	1	-	-
Monoalkyl	2	-	-
Dialkyl	3	-	-
Trialkyl	3	-	2
Acyl	1	-	-
Bicyclic	2	1	1
2-Thiazolines	1	-	1
3-Thiazolines	2	-	-
Other	1	-	-
Total	16	1	4
% of total volatiles	2	1	2

On the basis of percentage of total volatiles, beef and lamb
show an equal figure. However inspection of the total number
shows that beef has a far larger number of thiazoles and
thiazolines than lamb does. Only in the sub-groups bicyclic,
trialkyl and 2-thiazolines is there a comparable number of
compounds, especially between beef and lamb. Benzothiazole is
common to all three species and 2,4,5-trimethyl thiazole and 2,4-
dimethyl-5-ethyl thiazole (25) are common to beef and lamb. This
particular thiazole has been reported as having a Buchu leaf oil
aroma and a meaty liver flavour[75] and it has been suggested that

it could be a major contributor to the aroma of roast beef.[84] Beef has several thiazoles and thiazolines unique to its own species including 2-acetyl thiazole which is reported as having a bready, burnt, nutty, cereal, popcorn type aroma. Another unique thiazole is 4-methyl-5-(2-hydroxyethyl) thiazole (18), which is a primary degradation product of thiamine (as previously described). Beef also has a large number of alkyl substituted thiazoles which range from green and vegetable-like aromas at the lower end to nutty, roasty and meaty as the substitution increases.[3] Pittet and Hruza[75] observed that substitution in the 2- or 4-position adjacent to the ring nitrogen resulted in more potent and characteristic flavours than substitution in the 5-position. Certainly the thiazoles and thiazolines appear to have much lower odour thresholds than the oxazoles and oxazolines and therefore probably contribute much more to the overall aroma of beef.

To understand the mechanisms for the production of thiazoles in the meat cooking process, we have to consider the basic production of hydrogen sulphide. Thiazoles are not primary flavour components but rather a combination or interaction of a variety of primary products from the Maillard reaction. Hydrogen sulphide is all-important in the proposed pathways for thiazole formation and is formed during the Strecker degradation of cysteine (Fig.6) and from other sulphur amino acids and thiamine.

There are a number of routes proposed for the formation of thiazoles. Already mentioned is the degradation of thiamine to produce 4- and 5-substituted thiazoles. Also reported is the reaction of cysteamine, produced from cysteine, with aldehydes, produced by the Strecker degradation of amino acids to give condensation and ring closure. However, by far the more frequent are the reactions of α-dicarbonyl compounds, produced from the degradation of sugars, with hydrogen sulphide, ammonia and an aldehyde.

Fig.14 shows that the key intermediates in this pathway are α-mercaptoketones formed by the nucleophilic attack of hydrogen sulphide on dicarbonyls. The intermediate mercaptoketones then react with imines formed from the interaction of aldehydes with ammonia, to produce thiazolines which go on to yield the final thiazole products. The example used in Fig.14 is the formation of 2,4-dimethyl-5-ethyl-thiazole (25), chosen because of its potential importance in the aroma of roast beef.

Considering the various pathways proposed for the formation
of thiazoles, it can be seen that all the reactants are
intermediates formed from the primary precursors - fats, proteins
and carbohydrates.   Lamb, it is reported, has a very high
cysteine and cystine content[79] and also, because of high fat
content, should form aldehydes readily.   Furthermore, water
extracts of lamb muscle are high in glutathione[85] thus creating a
situation where aldehyde/hydrogen sulphide reactions should
predominate. This is observed in part because a high proportion
of lamb volatiles have been identified as cyclic sulphur systems
and this is reported later, but there are few thiazoles and
thiazolines. This can be explained by the fact that lamb muscle
is much lower in carbohydrate content[60] and so the level of
diketones, critical to the formation of thiazoles, will be
greatly reduced.   As described earlier, pork contains a high
level of thiamine and also has 1.2-1.3 times as much hydrogen
sulphide as cooked beef.   One would therefore expect that with
further work a number of thiazole systems await to be detected in
this meat.

Non-cyclic sulphur compounds

The importance of sulphur compounds both aliphatic and

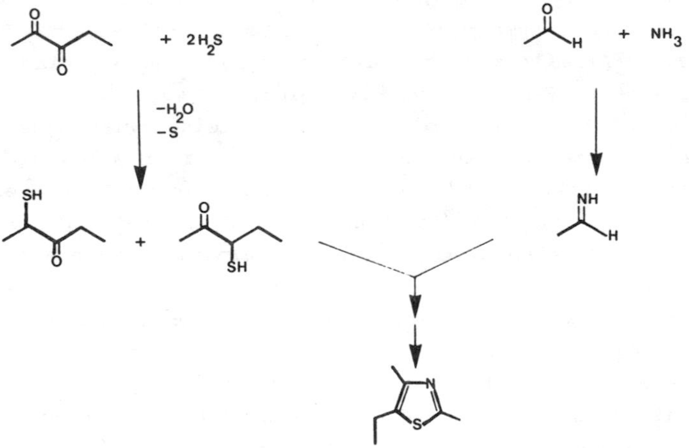

(25)

Figure 14 Formation of thiazoles

cyclic in the overall aroma of cooked meat has been stressed by many groups of researchers. Wasserman[5] emphasised that when sulphur compounds were removed from meat preparations, reduction or even elimination of meaty notes was observed. This section will deal with the aliphatic thiols, sulphides, polysulphides and other non-cyclic sulphur compounds; in other words, all the compounds which contain sulphur but not incorporated in a cyclic system.

Looking at Table 15, it can be seen that beef contains by far the greatest number of such sulphur compounds, although pork in fact contains the same on a percentage basis. Pork and lamb contain no aromatic sulphides or mercaptans and this group may be an important contributor to the overall aroma of roast beef. The mercaptan category reveals that beef has 5 dithiols, lamb contains one and pork has none. Although the sulphides and disulphides are represented in each species, beef has a far wider range of these types of compounds. Lamb has no trisulphides reported but pork and beef each have three representatives in this category. Beef is the only species to have a tetrasulphide identified in it. A novel compound has been identified in roasted pork meat headspace and this compound, 4,6-dimethyl-2,3,5,7-tetrathiaoctane[36] (27) so far has not been reported as being present in either of the other two species. 1-(Methylthio)ethanethiol (26) has been reported in beef[86] but

Table 15 Non-cyclic sulphur compounds in the red meats

	Beef	Pork	Lamb
Aromatic sulphides and mercaptans	9	–	–
Aliphatic sulphides	19	2	2
Aliphatic polysulphides	17	6	1
Mercaptans	20	8	1
Thioesters	3	5	–
Others	5	5	2
Total	73	26	6
% of total volatiles	11	11	3

not in either lamb or pork. The other types of compounds present
are thioesters, found in pork and beef but not in lamb.

With regard to organoleptic properties, great interest has
been shown in many of the sulphur compounds falling within the
categories described in this section. It was felt that these
compounds could be key contributors to the aroma of cooked meat
flavours, but in general terms they possess non-meaty,
objectionable odours. Sulphides and polysulphides are
characterised by strong onion, garlic-like aromas. Thiols, as
reported by Maga,[87] display unpleasant odours described as
sulphurous, cabbage-like, garlic-like and oniony. However, there
are exceptions reported in the literature. 2-Methylbenzene
thiol, found in beef by Garbusov et al.[88] is reported to have a
meat broth-like aroma. Other 2-substituted benzene thiols,
although not reported in meat, may also play a role in beef
aroma. Methyl butanethiols[89] have a meaty type, stewy note at
very low levels and are incorporated as flavouring agents in
synthetic meat flavours. To complete the thiol group a close
examination of the dithiols, particularly the α, ω-type (i.e.
$HS-(CH_2)_n-SH$), shows that as n increases, the organoleptic
character changes from oniony, garlic-like through burnt, fatty
meaty (n = 6) to sweet fatty. Hexane-1,6-dithiol has been
identified in beef.[88] The dimercapto-butanes are also an
interesting group and 2,3-dimercapto-butane derived from
3-mercapto-2-butanone is reported to be an important meat
chemical.[4] In general terms, the odour thresholds of all these
above-mentioned compound types are very low and so it might be
expected that these play an important part in the overall aroma
of cooked meats.

The formation of these compounds comes from a fairly complex
network of pathways, some of which are outlined in Fig.15. The
common building blocks are hydrogen sulphide, methanethiol,
acetaldehyde and dimethyl disulphide, and all are reported in the
three species in significant quantities.

Hydrogen, sulphide and methanethiol are generated in large
quantities via the Strecker degradation of cysteine and
methionine respectively. Methanethiol can be oxidised to form
dimethyl disulphide which can directly form dimethyl sulphide and
dimethyl trisulphide. Thus initial fragmentation of sulphur-
containing amino acids followed by dimerisation and

polymerisation of the small sulphur-containing molecules produced can lead to a wide range of sulphur compounds. It is known that sulphur compounds in particular can be taken up by fat and retained there to a greater extent than in aqueous broth.[90] This would therefore allow the interaction of hydrogen sulphide and the other sulphur compounds with fatty aldehydes produced from lipid oxidation/degradation. It is reported that this can give primary and secondary reactions leading to a whole range of thiols, dithiols and sulphides.[91] Acetaldehyde, produced from the Strecker degradation of alanine, reacts with hydrogen sulphide to form the very reactive intermediate, ethane-1,1-dithiol (28) which can react with itself to produce bis-(1-mercaptoethyl) sulphide (29), a constituent of lamb meat. Fig.15 outlines the reaction pathway.

1-(Methylthio)ethanethiol (26), another important precursor for various sulphur systems, can be formed by the interaction of acetaldehyde and methanethiol. Dubs and Stussi[92] proposed that 1-(methylthio)ethanethiol (26) can further interact with dimethyl disulphide to produce compounds of the tetrathiaoctane type (27)

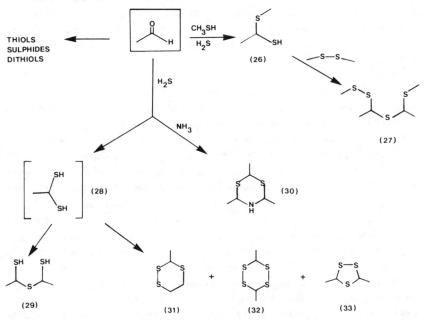

Figure 15 Formation pathways for sulphur systems

identified in roast pork.

Cyclic sulphur compounds

Cyclic sulphur compounds, excluding those heterocyclic classes previously discussed, follow on very closely from the non-cyclic sulphur compounds and the actual numbers found in each species are:

Beef - 15 (2%)   Pork - 4 (2%)   Lamb - 5 (2%)

3,5-Dimethyl-1,2,4-trithiolane (33) and ethylene sulphide are common to all three species, and thialdine (30) is common to beef and lamb. It should be noted that Nixon et al.[93] observed that thialdine was present at a concentration greater than 20% of total volatiles and 3,5-dimethyl-1,2,4-trithiolane was present at between 10 and 20% of a lamb extract. It is also conceivable that ring systems containing more than one nitrogen, i.e. thiadiazines, will also be present. Beef contains a series of dithianes, trithianes and trithiolanes with a variety of alkyl substituents, all unique to that species. Particularly worthy of note is 3-methyl-1,2,4-trithiane (31), subject of a patent,[94] and used in meat flavourings because of its low aroma threshold and high potency. Pork contains a unique compound, 1,3-thioxalane, and the compounds 3,6-dimethyl-1,2,4,5-tetrathiane (32) and 1,2,3,5,6-pentathiepane (lenthionine) have been found only in lamb.

The organoleptic properties of these systems tend to be sulphury floral, musty, onion-like, and are thought to embellish the stew-like note of cooked meats.

The formation of these compounds is very closely related to the pathways described for the non-cyclic sulphur compounds (Fig.15) previously discussed. Acetaldehyde, on reacting with hydrogen sulphide, can quite easily be shown to produce a range of cyclic compounds containing sulphur, the prime examples being the trithiolanes, the trithianes, tetrathianes and, if ammonia is introduced, thialdine (30). In effect we have a common starting point because the reaction of acetaldehyde with hydrogen sulphide and/or methanethiol was shown to produce a range of non-cyclic sulphur compounds which, upon further treatment, could themselves be cyclised. An example would be bis-(1-mercaptoethyl) sulphide (29) which could quite easily cyclise to form 3,5-dimethyl-1,2,4-trithiolane (33). It is easy to see that this whole area of sulphur chemistry is interlinked and interdependent, the net result being the formation of a wide range of potent flavour chemicals as a result of the secondary reactions of reactive intermediates produced by the breakdown of primary precursors.

The interaction of aldehydes with sulphur compounds is therefore a very important process in meat flavour development. However, it has been shown that lamb contains a higher concentration of sulphur amino acids[85] than both beef and pork and has quite a high fat content potentially leading to the formation of a considerable amount of short chain aldehydes. The sections on sulphur compounds in this chapter show that beef contains much greater numbers, but this is because many workers have specifically looked for sulphur compounds using selective extraction techniques and specific detection systems. With respect to lamb, copious black precipitates have been observed in mercuric cyanide collection traps when the volatiles were passed through, implying a high concentration of sulphur compounds.[95] Nixon et al.[93] also report that hydrogen sulphide was detected in lamb volatiles at a concentration greater than 20%. Obviously a greater number of sulphur compounds await detection in lamb.

Miscellaneous

This section deals with a variety of classes not covered under the major sections already described. Some of these classes are only minor contributors on a percentage basis whilst others contain a relatively large number of compounds. The hydrocarbons for example (Table 16), are the largest individual group of compounds identified in beef.

Table 16 Hydrocarbons in the red meats

	Beef	Pork	Lamb
Saturated straight chain	23	12	13
Saturated branched chain	18	–	–
Unsaturated aliphatic	27	1	–
Aromatic	43	17	16
Others	9	–	4
Total	120	30	33
% of total volatiles	18	13	14

Again, the combination of high temperature roasting methods favouring hydrocarbon formation and more workers investigating this species account for the greater numbers found in beef. Mottram et al.[31] observed that beef samples had much higher levels of 1,2-dimethylbenzene than did pork. Min et al.[96] have suggested that alkylbenzenes, which possess mothball, fatty, green aroma characteristics, could play a role in the odour of cooked beef fat. Lawrie[7] reported that low molecular weight hydrocarbons tended to result from underdone beef, whereas more intensive cooking produced high molecular weight hydrocarbons.

Alcohol classes identified in the red meats are outlined in Table 17. Not only have a greater number been identified in beef but quantitatively this meat contains more. Mottram et al.[31] reported that beef contains significantly larger amounts than

Table 17 Alcohols in the red meats

	Beef	Pork	Lamb
Straight chain aliphatic	17	10	8
Branched chain aliphatic	17	12	2
Unsaturated aliphatic	9	2	2
Aromatic	7	–	2
Alicyclic	2	1	–
Others	3	–	–
Total	55	25	14
% of total volatiles	8	11	6

pork of 1-pentanol, 1-octen-3-ol and 1-heptanol. Bodrero et al.[2] investigating the contribution of flavour volatiles to meat aroma using surface response methodology, concluded that unsaturated alcohols, especially 1-octen-3-ol, make a contribution to meat aroma.

MacLeod and Coppock[97] proposed the importance of pyrroles in the roasted aromas of heated beef. Organoleptically, pyrroles generally impart a burnt and earthy note, but 1-pyrroline, which has been found in lamb, gives a significantly enhanced butter flavour when added to manufactured margarine.[98] Apart from l-pyrolline, which is unique to lamb, the only other pyrroles identified in the three species have all been found in beef. Of these compounds the majority are ring alkyl substituted with 2-formyl and 2-acetyl pyrroles also contributing. Pyrroles are formed by the pyrolysis of amino acids such as proline, by the Maillard interaction of sugars and amino acids such as proline and glucose, and also by the interaction of furans and ammonia, the oxygen of the furan being exchanged by the nitrogen of the ammonia.

A recently reported class of meat volatiles are the pyrimidines and here beef has two representatives whereas none has been reported in either pork or lamb. The pyrimidines found in beef are 4,6-dimethylpyrimidine (34) and 4-acetyl-2-methyl-pyrimidine (35), and the latter is reported as having an interesting roasted note and being detectable at 0.5 ppm.[48]

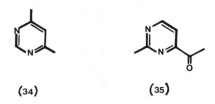

(34)   (35)

CONCLUSIONS

Apart from the contribution made by taste components such as nucleotides, free amino acids and lipids, the flavour of meat is formed during the cooking process from a range of primary precursors which react to furnish a multitude of aroma volatiles. Some of these aroma volatiles either contribute directly to meat flavour or form a reactive pool of secondary intermediates such as aldehydes, diketones, hydrogen sulphide, ammonia, etc., which interact to produce the final flavour complex (Fig.16). The key compound in the whole of this process is hydrogen sulphide, which makes its own flavour contribution but, more importantly, plays a critical role in the formation of many flavour chemicals which define the character of meat.

Depending upon the species and the cooking process involved, the secondary reactive pool of intermediates will vary and hence the character of the final meat flavour will vary. Considering the different meat species, the flavour of lamb is influenced by a high pyridine content, by the presence of branched chain fatty acids and by a high concentration of cyclic sulphur compounds. Thus amino acids, especially the sulphur-containing amino acids, and lipids have a major influence on the flavour of lamb. Pork flavour is also strongly influenced by the lipid content, but in the case of this meat, thiamine and its degradation products play an important role. In beef the fat content plays a much less prominent role and does not have a dominating flavour effect over the water-soluble flavour-forming reactions occurring in the lean. Thus, the characteristic flavour of the total Maillard process and the interaction of hydrogen sulphide with compounds such as hydroxy furanones contribute the major part of the flavour of this meat. In roasted beef the flavour character is influenced to a greater extent by compounds such as thiazoles and pyrazines formed more readily at higher cooking temperatures.

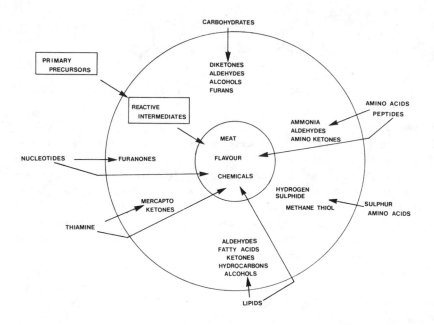

Figure 16 The flavour of meat - a summary

In terms of consumer preferences for the species meats, probably the most important factor is the influence and quality of the lipid fraction. Lamb possesses a hard saturated fat which is unpleasant when cold and contributes some unusual branched chain acids to the flavour. Pork contains a highly unsaturated fat which, although much more acceptable in flavour terms, is prone to oxidative attack leading to the formation of a high content of lipid derived volatiles. Only in beef is the highly desirable flavour character of lean meat allowed to dominate, relatively unhindered by contributions from the fat, and this fact alone probably accounts for its popularity over the other two species in the Western world.

ACKNOWLEDGEMENTS

The authors acknowledge the contribution made to this document by Dr N.W.R. Daniels and Mr R. Fensom of Dalgety Spillers Research and Technology Centre, Cambridge.

DEDICATION

This paper is dedicated to the memory of the late Willie Parker, Professor of Chemistry at the University of Stirling, Scotland, for whom both authors worked for a short period in their careers.

REFERENCES

1. C.G. May and I.D. Morton, British Patent 858,660, 1961.
2. K.O. Bodrero, A.M. Pearson and W.T. Magee, J.Food Sci., 1981, 46, 26.
3. G. MacLeod and M. Seyyedain-Ardebilli, Crit.Revs.Food Sci.Nutn., 1981, 14, 309.
4. I. Katz, "Flavour Research - Recent Advances", Marcel Dekker, New York, 1981, Ch.7, p.217.
5. A.E. Wasserman, J.Food Sci., 1979, 44, 6.
6. T. Shibamoto, J.Agric.Food Chem., 1980, 28, 237.
7. R.A. Lawrie, Food, 1982, 4, (2), 11.
8. A. Kuninaka, "Flavour Research - Recent Advances", Marcel Dekker, New York, 1981, Ch.9, p.305.
9. J.C. Boudreau, J. Ooravec, N.K. Hoang and T.D. White, "Food Taste Chemistry", ACS Symposium Series, 115, Washington D.C., 1979, p.1.

10. A Kuninaka, "The Chemistry and Physiology of Flavours", Avi Publishing Co. Inc., Westport, Connecticut, 1967, Ch.24, p.515.

11. A. Kuninaka, J.Agr.Chem.Soc.Japan, 1960, 34, 489.

12. A.F. Mabrouk, "Phenolic, Sulphur and Nitrogen Compounds in Food Flavours", ACS Symposium Series, 26, Washington D.C., 1976, Ch.10, p.146.

13. M. Motono, "Chemistry of Foods and Beverages: Recent Developments", Academic Press, New York, 1982, p.181.

14. Y. Komata, "Microbial Production of Nucleic Acid Related Substances", John Wiley & Sons, New York, 1976, p.299.

15. T. Hosoi, Protein, Nucleic Acid, Enzyme, 1961, 6, 115.

16. B. Toi, S. Maeda and H. Furukawa, Meeting of the Kanto Division of Agricultural Chemical Society of Japan, Tokyo, 12th November, 1960.

17. E.K. Fujita, B.S. Yasumatsu and Y. Uda, Meeting of the Kanto Division of Agricultural Chemical Society of Japan, Fukuoka, 1st April, 1961.

18. Kyowa Fermentation Industry Co. Ltd., Japanese Patent 55,377, 1965.

19. Kikkoman Shoyu Co.Ltd., Japanese Patent 15,743, 1967.

20. Kikkoman Shoyu Co.Ltd., French Patent 1,469,024, 1967.

21. Kyowa Fermentation Industry Co.Ltd., Belgian Patent 666,087, 1965.

22. C.H. Kurtzman and L.B. Sjostrom, Food Tech., 1964, 18, 1467.

23. E. Kirova and M. Peschevska, Die Nahrung, 1982, 26, 669.

24. D.J. Schinneller, R.H. Dougherty and R.H. Briggs, J.Food Sci., 1972, 37, 935.

25. Y. Yamasaki and K. Maekawa, Agric.Biol.Chem., 1978, 42, 1761.

26. Y. Yamasaki and K. Maekawa, Agric.Biol.Chem., 1980, 44, 93.

27. L. Schutte, "Phenolic Sulphur and Nitrogen Compounds in Food Flavours", ACS Symposium Series, 26, Washington D.C., 1976, Ch.6, p.96.

28. C. Hsu, R.J. Peterson, Q.Z. Jin, C. Ho and S.S. Chang, J.Food Sci., 1982, 47, 2068.

29. R.V. Golovnja and M. Rothe, Die Nahrung, 1980, 24, 141.

30. V.P. Uralets and R.V. Golovnja, Die Nahrung, 1980, 24, 155.

31. D.S. Mottram, R.A. Edwards and H.J.H. MacFie, J.Sci.Food

Agric., 1982, 33, 934.

32. K.N. Lee, C. Ho, C.S. Giorlando, R.J. Peterson and S.S. Chang, J.Agric.Food Chem., 1981, 29, 834.

33. "Volatile Compounds in Food", edited by S. van Straten, Division for Nutrition and Food Research TNO, Institute CIVO, Netherlands, 4th Edn, 1977, Suppl.7, p.61.

34. K. Lundstrom, K.-E. Hansson, S. Fjelkner-Modig and J. Persson, "Proceeding of the 26th European Meeting of Meat Research Workers", American Meat Science Association, Chicago, Illinois, 1980, 1, p.300.

35. R.V. Golovnya, V.G. Garbuzov, I. Ya. Grigoryeva, S.L. Zharich and A.S. Bolshakov, Die Nahrung, 1982, 26, 89.

36. P. Dubs and R. Stussi, Helv.Chim.Acta, 1978, 61, 2351.

37. J.R. Champagne and W.W. Nawar, J.Food Sci., 1969, 34, 335.

38. "Volatile Compounds in Food", edited by S. van Straten, Division for Nutrition and Food Research TNO, Institute CIVO, Netherlands, 4th Edn, 1977, Suppl.4, p.60.

39. G. Hoffmann and P.W. Meijboom, J.Amer.Oil Chem.Soc., 1968, 45, 468.

40. I. Hornstein and P.F. Crowe, J.Agric.Food Chem., 1960, 8, 494.

41. I. Hornstein and P.F. Crowe, J.Agric.Food Chem., 1963, 11, 147.

42. E. Wong, L.N. Nixon and C.B. Johnson, J.Agric.Food Chem., 1975, 23, 495.

43. G.A. Garton, F.D.De B. Hovell and W.R.H. Duncan, Br.J.Nutr., 1972, 28, 409.

44. G.A.M. Van den Ouweland, Perfumer & Flavourist, 1980, 5, 15.

45. O.F. Batzer, A.T. Santoro, M.C. Tan, W.A. Landmann and B.S. Schweight, J.Agric.Food Chem., 1960, 8, 498.

46. A.A. Wasserman and A.M. Spinelli, J.Agric.Food Chem., 1972, 20, 171.

47. E. Kevei and E. Kozma, Die Nahrung, 1976, 20, 243.

48. I. Flament, B. Willhalm and G. Ohloff, "Flavor of Foods and Beverages", Academic Press, New York, 1978, p.15.

49. R.J. Park, K.E. Murray and G. Stanley, Chem.Ind., 1974, 380.

50. H.T. Badings, Ned.Melk-Zuiveltijdschr., 1970, 24, 147.

51. M. Loury, Lipids, 1972, 7, 671.

52. W. Grosch, "Food Flavours", Elsevier, Oxford, 1982, Ch.5, p.325.

53. F. Caporaso, J.D. Sink, P.S. Dimick, C.J. Mussinan and A. Sanderson, J.Agric.Food Chem., 1977, 25, 1230.

54. W. Pickenhagen, British Patent 2,031,261, 1980.

55. T. Yamato, T. Kurata, H. Kato and M. Fujimaki, Agric.Biol. Chem., 1970, 34, 88.

56. S.S. Chang and R.J. Peterson, J.Food Sci., 1977, 42, 298.

57. R.L.S. Patterson, J.Sci.Food Agric., 1968, 19, 31.

58. T. Shibamoto and G.F. Russell, J.Agric.Food Chem., 1976, 24, 843.

59. T. Shibamoto and G.F. Russell, J.Agric.Food Chem., 1977, 25, 109.

60. R.L. Macy, H.D. Naumann and M.E. Bailey, J.Food Sci., 1964, 29, 142.

61. A.A. Paul and D.A.T. Southgate, "McCance and Widdowson's The Composition of Foods", Her Majesty's Stationery Office, London, 4th Rev.Edn, 1978.

62. S.S. Chang, T.H. Smouse, R.G. Krishnamurthy, B.D. Mookherjee and R.B. Reddy, Chem.Ind., 1966, 1926.

63. C.H.T. Tonsbeek, A.J. Plancken and T. Von der Weerdof, J.Agric.Food Chem., 1968, 16, 1016.

64. G.A. Van den Ouweland and H.G. Peer, J.Agric.Food Chem., 1975, 23, 501.

65. N.J. Evers, H.H. Heinsohn, B.J. Mayers and A. Sanderson, "Phenolic, Sulphur and Nitrogen Compounds in Food Flavours", ACS Symposium Series, 26, Washington D.C., 1976, Ch.11, p.184.

66. B.K. Dwivedi and R.G. Arnold, J.Agric.Food Chem., 1973, 21, 54.

67. L.M. Van der Linde, J.M. Van Dort, P. de Valois, M. Boelens and D. de Rijke. "Progress in Flavour Research", Applied Sciences, London, 1979, p.219.

68. R.M. Seifert, R.G. Buttery, R.W. Lundin, W.F. Haddon and M. Benson, J.Agric.Food Chem., 1978, 26, 1173.

69. R.G. Buttery and R.M. Seifert, J.Agric.Food Chem., 1982, 30, 1261.

70. R.G. Buttery, R.M. Seifert, J.G. Turnbaugh, D.G. Guadagni and L.C. Ling, J.Agric.Food Chem., 1981, 29, 183.

71. J.A. Maga and C.E. Sizer, "Fenaroli's Handbook of Flavour Ingredients", CRC Press, Cleveland, Ohio, 2nd Edn, 1975, 1, 47.

72. I. Flament, "Proc. 2nd International Conf. Flavs. and Fragrances - Athens", Academic Press, 1981, 1, 35.

73. C.J. Mussinan, R.A. Wilson and I. Katz, J.Agric.Food Chem., 1973, 21, 871.

74. I. Flament, M. Kohler and R. Aschiero, Helv.Chim.Acta, 1976, 59, 2308.

75. A.O. Pittet and D.E. Hruza, J.Agric.Food Chem., 1974, 22, 264.

76. J.A. Newell, M.E. Mason and R.S. Matlock, J.Agric.Food Chem., 1967, 15, 767.

77. M. van Praag, H.S. Stein and M.S. Tibbetts, J.Agric.Food Chem., 1968, 16, 1005.

78. P.E. Koehler, M.E. Mason and J.A. Newell, J.Agric.Food Chem., 1969, 17, 393.

79. R.G. Buttery, L.C. Ling, R. Teranishi and T.R. Mon, J.Agric.Food Chem., 1977, 25, 1227.

80. Y.C. Lien and W.W. Nawar, J.Food Sci., 1974, 39, 914.

81. K. Suyama and S. Adachi, J.Agric.Food Chem., 1980, 28, 546.

82. C.J. Mussinan, R.A. Wilson, I. Katz, A. Hruza and M. Vock, "Phenolic, Sulphur and Nitrogen Compounds in Food Flavours", ACS Symposium Series, 26, Washington D.C., 1976, p.133.

83. G. Vernin and C. Parkanyi, "The Chemistry of Heterocyclic Flavouring and Aroma Compounds", Ellis Horwood, Chichester, 1982, Ch.3, p.181.

84. A.M. Galt, Ph.D. Thesis, Queen Elizabeth College, University of London, 1981.

85. R.L. Macy, Jnr., H.D.N. Naumann and M.E. Bailey, J.Food Sci., 1964, 29, 136.

86. H.W. Brinkman, H. Copier, J.J.M. de Leuw and S.B. Tjan, J.Agric.Food Chem., 1972, 20, 177.

87. J.A. Maga, CRC Crit.Revs.Food Sci.Nutr., 1976, 7, 147.

88. V. Garbusov, G. Rehefeld, G. Wolm, R.V. Golovnja and M. Rothe, Die Nahrung, 1976, 20, 235.

89. R.A. Wilson, C.J. Mussinan, I. Katz and A. Sanderson, J.Agric.Food Chem., 1973, 21, 873.

90. K.O. Herz and S.S. Chang, Adv.Food Res., 1970, 18, 1.

91. M. Boelens, L.M. van der Linde, P.J. de Valois, H.M. van Dort and H.J. Takken, J.Agric.Food Chem., 1974, 22, 1071.

92. P. Dubs and R. Stussi, Helv.Chim.Acta, 1978, 61, 2355.

93. L.N. Nixon, E. Wong, C.B. Johnson and E.J. Birch, J.Agric.

Food Chem., 1979, <u>27</u>, 355.

94. I. Flament, <u>British Patent</u> 2,010,263 A, 1979.

95. D.A. Cramer, <u>New Mex.State Univ.Exp.Sta.Bull.</u>, 1974, <u>616</u>, 24.

96. D.B.S. Min, K. Ina, R.J. Peterson and S.S. Chang, <u>J.Food Sci.</u>, 1977, <u>42</u>, 503.

97. G. MacLeod and B.M. Coppock, <u>J.Agric.Food Chem.</u>, 1977, <u>25</u>, 113.

98. G.M. Nakel, <u>U.S. Patent</u> 3,336,138, 1967.

8
Fundamental Radiation Chemistry of Food Components

By A. John Swallow
PATTERSON LABORATORIES, CHRISTIE HOSPITAL AND HOLT RADIUM INSTITUTE,
MANCHESTER, U.K.

INTRODUCTION

Treatment of food with high-energy radiation can reduce or eliminate several of the undesirable changes which may take place during storage, including spoilage by insects and micro-organisms, and biochemical changes which lead to processes such as sprouting of potatoes. Irradiation has several advantages over other methods of treatment. For example the food can be processed cold, inside a sealed package if desired, and by a continuous rather than a batch process. For these and other reasons there has been an interest in food irradiation for more than half a century.

When considering the irradiation of food, it is necessary to enquire into any undesirable effects of irradiation. Foremost among these, especially in the lay mind, is the production of radioactivity. This could occur with radiations such as α-particles, neutrons or excessively energetic γ- or X-rays (>5MeV) or electrons (>10MeV). However the radiations of practical interest do not produce significant radioactivity.[1] These radiations include γ-rays with energies in the region of 1MeV, X-rays with energies less than 5MeV and fast electrons with energies below 10MeV. Although radioactivity is not a problem, chemical changes occur which may or may not be desirable. This paper discusses the basic processes taking place.

INTERACTION OF RADIATION WITH MATTER

γ-Rays and X-rays interact with matter by a process known as Compton scattering in which a photon dissipates some of its energy by ejecting an electron from a molecule before carrying on with reduced energy in a different direction. Fig.1 illustrates the process. Some of the energy departs from the sample in the form of the energy of the scattered photons, but some of it

remains, initially in the form of kinetic energy of the ejected
electrons.   The fraction remaining depends on the scattering
angle, but in a typical Compton scattering event is about half
the   incident   energy.   Another   process   is   photoelectric
absorption, in which the photon is totally absorbed by an atom,
its energy being used to eject an electron from an inner shell.
This process accounts for a significant proportion of the
interaction of low energy photons (less than about 100keV) with
matter of the type found in foodstuffs.   Pair production is a
process in which the energy of the photon is used to create a
positron-electron pair within the medium.   This process accounts
for a small proportion of the interactions of more energetic
photons (energy >1MeV) with the type of matter found in
foodstuffs.

Fast electrons, whether produced by γ- or X-irradiation or
directly from an electron accelerator, lose energy by exciting
and ionising molecules along their track at random.   The length
of the track varies from a fraction of a millimetre for some of
the electrons produced by γ- or X-rays to several centimetres for
10MeV electrons.   Some of the electrons ejected in the ionisation

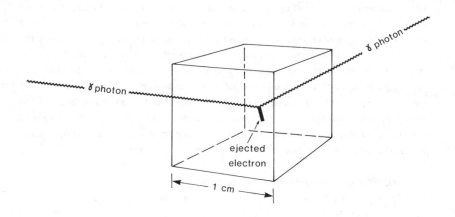

Figure 1   Typical Compton scattering event produced by a γ-ray
photon. 10^{12}-10^{14} such   events   would   occur   in   each   cubic
centimetre during radiation treatment.

event have sufficient energy to cause further excitations and ionisations. If the energy enables the electron to depart far from its point of origin, a branched track is produced. However it is more common for the energy to be quite low, in which case a small number of further excitations and ionisations are produced in a 'spur' along the main track. An illustration of a typical electron track is shown in Fig.2. Positrons lose energy in the same way as negatively charged electrons, and at the end of their track recombine with electrons, the energy liberated appearing in the form of photons of energy 0.51MeV, which may escape from the sample or may in turn interact by Compton scattering or photoelectric absorption.

The energy expended in exciting or ionising a molecule is in the region of 10-50eV. Consequently each fast electron will excite or ionise tens or hundreds of thousands of molecules. Comparing the effect of Compton scattering or photoelectric absorption events as such with the effect of the Compton or photo-electrons produced in the events, it is clear that the Compton scattering or photoelectric absorption events themselves must account for only a trivial proportion of the effects. In other words, whether matter is irradiated with γ- or X-rays or fast electrons, it is excitations and ionisations produced by fast electrons which are the important source of chemical change.

The interaction of high-energy radiations with matter can be contrasted with the interaction of ultra-violet or visible light. There are two significant differences. One is that

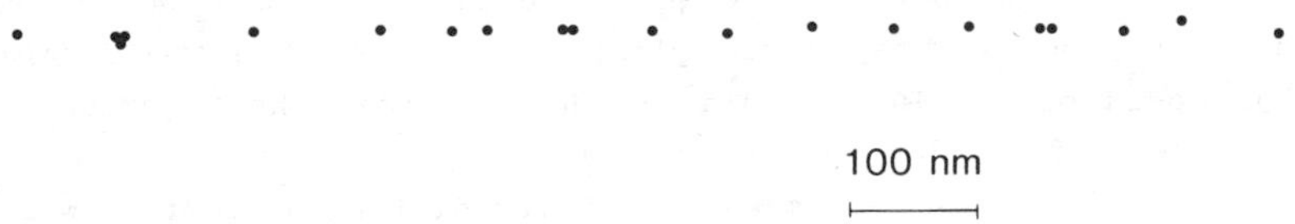

100 nm

Figure 2   Positive ions along part of typical electron track such as produced by γ-, X- or electron irradiation. One electron is produced for every positive ion. The distribution of excited molecules would be similar.

high-energy radiations produce different intermediates, i.e. excited states which generally have higher energy than those produced photochemically, together with positive ions and the associated electrons. The other is that high-energy radiations are able to act randomly on all of the molecules in the medium whereas light can be absorbed only by molecules possessing an appropriate chromophore.

RADIATION DOSIMETRY

The extent of any change produced by irradiation must depend on the amount of radiation energy deposited in the sample. The measure of this quantity is the radiation dose, the unit of which is the gray (Gy), equal to $1 Jkg^{-1}$. An older unit, the rad, is still found in the literature. 1 rad is one hundredth of a gray. For purposes of calculation it is sometimes convenient to express doses in terms of eV, where $1eV = 1.6 \times 10^{-19}J$:

$$1Gy = 1Jkg^{-1} = 100 \text{ rad} = 6.24 \times 10^{18} eVkg^{-1}$$

Most of the food irradiation processes under consideration require doses in the region of 0.1-10kGy.

The unit used in discussions of the chemical effects of irradiation is the G-value, defined as the number of molecules changed per 100eV absorbed in the medium. The physical processes discussed above result in the first instance in around 3 molecules being excited or ionised per 100eV, i.e. G = 3.

If the whole of the radiation energy were degraded to heat, it can easily be shown that doses in the range 0.1-10kGy would produce temperature rises in the range 0.024-2.4°C. Similarly, if every molecule which was changed in the first instance resulted in one molecule being altered after the subsequent chemical processes had taken place, a G-value of 3 would imply that doses of 0.1-10kGy would produce chemical change amounting to a total of $3 \times 10^{-5} - 3 \times 10^{-3} mol \ dm^{-3}$.

Radiation doses are measured by irradiating dosimeters which provide a measurable response to radiation, and then using the dose in the dosimeter to calculate the dose which would be received by the sample of interest. The most fundamental dosimeters are calorimeters, but owing to the small temperature rises these are unsuitable for routine use. Ionisation chambers are also classical dosimeters employed in calibrating laboratories, but for practical purposes it is more convenient to use dosimeters such as chemical solutions or polymers which respond to radiation in a way that can be measured easily in an ordinary laboratory, and which have in turn been calibrated against standards.

One such system consists of an aqueous solution of ferrous ions in 0.4M sulphuric acid (Fricke dosimeter). When this solution is subjected to γ-, X- or electron radiation, close to 15.6 ferrous ions are converted to ferric per 100eV absorbed in the solution, so that measurement of ferric ions (using, for example, their absorption of UV light of wavelength 304nm) enables the dose to be calculated. Another dosimeter consists of crystals such as lithium fluoride which emit thermoluminescence on heating after irradiation. Pieces of polymethylmethacrylate, either with or without added dyes, are also widely used. There are also many, many others.

It is less easy than might be thought to calculate the dose in the sample from the dose in the dosimeter. It is also important to take into account any variations in dose throughout the sample, since any part of the food receiving less than the average dose may be insufficiently treated, while any part receiving more than the average could deteriorate in quality. Specialised publications may be consulted for further details (e.g. Chadwick et al.[2]).

RADIATION EFFECTS MEDIATED BY WATER

Most foods contain a high proportion of water. Since the action of radiation is random, a corresponding proportion of the interactions occurs in the first instance on water molecules. It is therefore necessary to consider the reactive intermediates formed by irradiation of water and the effects produced when these species attack other components. There is no evidence that the excited water molecules play an important part (the energy of excitation is probably simply converted to heat), but there are important consequences of the ionisation.

In chemical terms, the ionisation and its immediate consequences may be represented by the equations:

$$H_2O \longrightarrow H_2O^+ + e^- \qquad (1)$$

$$H_2O^+ + H_2O \longrightarrow H_3O^+ + {}^\bullet OH \qquad (2)$$

$$e^- + aq \longrightarrow e^-_{aq} \qquad (3)$$

Reactions 2 and 3 take place within picoseconds of the ionisation process, reaction 1. In regions of the track where two or more ionisations occur together (see Fig.2), additional processes occur including:

$$e^-_{aq} + H_3O^+ \longrightarrow H_2O + {}^\bullet H \qquad (4)$$

$$2\,{}^\bullet OH \longrightarrow H_2O_2 \qquad (5)$$

$$2e^-_{aq} + 2H_2O \longrightarrow H_2 + 2OH^- \qquad (6)$$

The net result of all these processes is that water gives rise to free hydroxyl radicals, hydrated electrons and hydrogen atoms, all of which are highly reactive, together with H_2, H_2O_2, H_3O^+ and OH^-. Experiments on the γ-, X- or electron irradiation of carefully chosen aqueous solutions have shown that in pure water the yields of the species of interest are close to those shown in Table 1. In a system like a foodstuff which contains a considerable proportion of matter other than water, the yields of the most active species would be about 10% higher than those in the Table because of various reactions taking place in the spurs.

Subsequent reactions are straightforward in the case of simple, well-defined systems such as the acid solutions of ferrous ions used in the Fricke dosimeter. In this system the G-values for primary products are close to ${}^\bullet OH = 2.9$, $e^-_{aq} = 3.2$, ${}^\bullet H = 0.5$ and $H_2O_2 = 0.8$. The hydroxyl radicals oxidise ferrous ions to ferric:

$$ {}^\bullet OH + Fe^{2+} \xrightarrow{\ H^+\ } H_2O + Fe^{3+} \qquad (7)$$

Table 1 Species formed by irradiation of pure water with γ-rays or fast electrons

$^{\bullet}OH$	$G = 2.7$
e^-_{aq}	2.7
$^{\bullet}H$	0.55
H_2	0.45
H_2O_2	0.71

The hydrated electrons react with the hydrogen ions (reaction 4) and hydrogen atoms react with the oxygen which is always present in solutions prepared in the presence of air:

$$^{\bullet}H + O_2 \longrightarrow {}^{\bullet}HO_2 \qquad (8)$$

The $^{\bullet}HO_2$ radical is the protonated form of the superoxide anion radical $O_2{}^{\bullet-}$. It oxidises ferrous ions to ferric, yielding also hydrogen peroxide:

$$^{\bullet}HO_2 + Fe^{2+} \xrightarrow{\ H^+\ } H_2O_2 + Fe^{3+} \qquad (9)$$

The hydrogen peroxide oxidises ferrous ions in the classical Fenton reaction, yielding also $^{\bullet}OH$ radicals which in turn react according to equation 7:

$$H_2O_2 + Fe^{2+} \xrightarrow{\ H^+\ } H_2O + {}^{\bullet}OH + Fe^{3+} \qquad (10)$$

The net result of reactions 7-10 is the formation of ferric ions with $G = 15.6$.

REACTIONS OF $^{\bullet}$OH AND $^{\bullet}$H

With organic compounds, a characteristic reaction of both $^{\bullet}$OH and $^{\bullet}$H is to abstract hydrogen atoms from C-H, N-H and S-H bonds. There is a distinct tendency to abstract from the weaker bonds. For example, attack on cysteine occurs almost exclusively at the sulphur atom even though the molecule contains three C-H bonds and two N-H bonds for every S-H.

Attack by $\cdot$OH is the dominant source of change when carbohydrates are irradiated in aqueous systems. In this case there is little selectivity of attack, glucose, for example, yielding six different carbohydrate radicals in similar yields. The carbohydrate radicals are able to react in three principal ways: elimination, mutual reaction and scavenging. The most common elimination suffered by the primary carbohydrate radicals is of water yielding acylalkyl radicals:

$$-\overset{\cdot}{C}OH-CHOH- \longrightarrow -CO-\overset{\cdot}{C}H- + H_2O \qquad (11)$$

Mutual reaction of the primary carbohydrate radicals is mainly by disproportionation, but the acylalkyl radicals are able to combine to yield dimeric materials as well as to abstract H from primary carbohydrate radicals. The primary carbohydrate radicals can react with scavengers such as certain oxidising transition metal ions or H_2O_2. When oxygen is present, they react with it to form peroxyl radicals, and these in turn react further. Acylalkyl radicals are oxidising agents. With polysaccharides, the glycosidic link appears especially susceptible to hydrolysis when there is a radical site at one of the carbon atoms linked. All the reactions have been elucidated in considerable detail by analysing the stable products formed on irradiation of pure solutions of carbohydrates under carefully chosen conditions.[3]

Another common reaction of both $\cdot$OH and $\cdot$H is to add to aromatic and olefinic groups. This is an important source of change in the irradiation of proteins, although significant abstraction of hydrogen from the peptide carbon-hydrogen bonds also takes place. Addition of $\cdot$OH may be followed by water elimination, as demonstrated with tyrosine.[4]

$$\cdot OH + \text{(aromatic ring structures)} + H_2O \qquad (12)$$

The radical centres formed at one point in a protein are able to transfer to another point within the same molecule. Thus electron-deficient radicals of tryptophan become converted with quite high efficiency into the electron-deficient phenoxyl radicals of tyrosine.[5] Some of the radicals possess reducing power, so that with proteins containing iron in the ferric state, attack by ·OH can result in quite efficient **reduction** of the iron, necessarily accompanied by an oxidative change in the protein. This is important in the irradiation of myoglobin. Other reactions of protein radicals result in dimerisations and disproportionations.[6]

Since a given dose of radiation produces a fixed number of radicals from water which then proceed to attack the other molecules in the system, it follows that the total number of molecules affected by a given dose is independent of the number of molecules present. If an enzyme were to be irradiated in dilute solution, the number of molecules which are changed sufficiently to cause the enzyme activity to be lost may represent a high proportion of the molecules present, so that a high degree of inactivation would result. If a concentrated solution of the same enzyme were to be irradiated with the same dose, the deactivated molecules would represent a smaller proportion of those present, so there would only be a small percentage of inactivation. The presence of molecules of a different type strongly diminishes the inactivation, by taking up some of the water radicals without transferring the effect to the enzyme. These phenomena have been known for several decades.[7] A consequence is that irradiation is not an effective method of inactivating the enzymes in foods.

REACTIONS OF e_{aq}^-

Hydrated electrons are very powerful reducing agents[8] (aq + e^- = e_{aq}^-, $E° = -2.87V$). They react, by simple addition in the first instance, with aromatic compounds, carbonyl compounds, sulphur-containing compounds and with carboxylic acids in their -COOH but not in their -COO⁻ forms. They add to oxygen forming the superoxide anion radical $O_2^{\cdot-}$. They do not react significantly with aliphatic hydrocarbons, alcohols or carbohydrates.

Attack by hydrated electrons is important in the irradiation of proteins in aqueous systems. The attack occurs in part at aromatic, sulphur-containing and other reactive groups, and in part at peptide bonds. Attack at sulphydryl groups is an important consideration in food irradiation, as H_2S is produced:

$$e^-_{aq} + RSH \longrightarrow R^{\cdot} + SH^-$$
$$\downarrow H^+ \qquad (13)$$
$$H_2S$$

Those electrons which react in the first instance at the peptide bond can give rise to deamination through a sequence of the following type:

$$e^-_{aq} + -NH-\overset{O}{\overset{\|}{C}}-CHR-NH_3^+ \longrightarrow -NH-\overset{O-}{\overset{|}{\underset{\cdot}{C}}}-CHR-NH_3^+ \qquad (14)$$

$$-NH-\overset{O-}{\underset{\cdot}{\overset{|}{C}}}-CHR-NH_3^+ \longrightarrow -NH-\overset{O}{\overset{\|}{C}}-\overset{\cdot}{C}HR + NH_3 \qquad (15)$$

Deamination requires the end amino group to be protonated. It appears to be possible even when the initial addition does not occur at a terminal peptide group, since the electron can be transferred along the polypeptide chain. Where deamination does not happen, the addition of an electron can be followed by chain scission:

$$-NH-\overset{O-}{\underset{\cdot}{\overset{|}{C}}}-CHR-NH-\overset{O}{\overset{\|}{C}}-CHR- \xrightarrow{H^+} -NH-\overset{O}{\overset{\|}{C}}-\overset{\cdot}{C}HR + NH_2-\overset{O}{\overset{\|}{C}}-CHR- \qquad (16)$$

The radical formed in this reaction can abstract a hydrogen atom from the peptide carbon-hydrogen bond to yield a radical of the same type as produced when $^{\cdot}OH$ attacks the polypeptide chain. These radicals can then dimerise or disproportionate.[9]

Intramolecular electron transfer after electron attack is sometimes very efficient. Thus almost 100% of the electrons which react with ferricytochrome-c end up by producing ferrocytochrome-c.[10] Similarly intermolecular electron transfers can take place very efficiently in mixtures, so that the initial reaction of hydrated electrons with one reducible compound may be followed by a sequence of electron transfers resulting in reduction of the most easily reduced compound present, even though its concentration is very small. Such processes enable minor components of foodstuffs (vitamins, colourants, etc.) to be significantly affected by irradiation even though few $^{\bullet}$OH radicals or e_{aq}^{-} attack them.

DIRECT ACTION

Direct action of radiation on a food component has some significance for major components like carbohydrates, proteins and lipids, where it adds to the effect produced by the radicals from water. When an electron is ejected from a molecule in a region of low dielectric constant, there is a high probability that it will become solvated or react with a solute while within the Coulomb field of its partner. The positive and negative species can then recombine, producing excited species which are sometimes sufficiently long-lived to have a chemistry of their own.

The direct ionisation of a molecule may be regarded as formally equivalent to attack by $^{\bullet}$OH radicals in that both processes represent electron loss or oxidation. Similarly, the attachment of the ejected electrons to other molecules is a reductive process formally equivalent to attack by hydrated electrons. The analogy with the action of $^{\bullet}$OH and e_{aq}^{-} should not, however, be pressed too far.

Direct action is especially important in lipids[11] where attack mediated by water must represent a smaller proportion of the total attack than with carbohydrates or proteins. From the very large variety of products identified (hydrocarbons, acids, products of higher molecular weight, etc.) it seems that the excitations and ionisations can give rise to bond scission at almost any point in the molecule. The electrons ejected in the ionisation process probably attach to carbonyl groups. From ESR studies, the carbonyl anion radicals so formed appear to be unstable except at temperatures greatly below 0°C, probably because of C-O scission according to:

$$-CH_2-O-\overset{\overset{\displaystyle O-}{|}}{\underset{\displaystyle \bullet}{C}}-R \longrightarrow -CH_2^{\bullet} + O = \overset{\overset{\displaystyle O-}{|}}{C}-R \qquad (17)$$

The radicals formed in this reaction may abstract hydrogen atoms from other molecules, a process which would be followed by further free radical reactions, or may link with other radicals forming stable products of higher molecular weight.[9]

The free radicals formed by irradiation can give rise to
peroxidation in the same way as radicals formed in more familiar
ways (see Chapter 6). Apart from the initiation step, similar
principles apply in the two cases. Hence unsaturated fats are
considerably more susceptible than saturated fats, and products
are formed in high yield as long as oxygen is present.
Antioxidants, metal ions, temperature, rate of initiation, etc.
would be expected to produce somewhat similar effects to those
produced in normal free radical autoxidation.

INFLUENCE OF IRRADIATION CONDITIONS

1.   Although the extent of change in food must depend on dose,
the response is not necessarily linear because minor components
may be so much affected in the earlier part of the irradiation
that subsequent effects become increasingly directed elsewhere.
Also, the irradiation products may themselves be attacked on
prolonged exposure.

2.   Most free radicals react with oxygen. It is usual for
oxidations to predominate as long as oxygen is present in the
sample (up to about 0.5-5kGy) depending on the type of food
(aqueous or fatty) and irradiation conditions. Once the oxygen
has gone, different reactions take place.

3.   Dose-rate affects response where there is competition
between radical-radical reactions (favoured by high dose-rates)
and radical-solute reactions (favoured by low dose-rates).
Dose-rate is especially important where there is a possibility of
free radical chain reactions taking place.

4.   A thin sample exposed to air may enable oxygen to diffuse in
while the irradiation is taking place, especially if the
dose-rate is low, whereas a thick sample exposed at a higher
dose-rate would be expected to be essentially anaerobic after
about a few kGy or less.

5.   The presence of added radical scavengers, antioxidants,
etc. can reduce the effects of irradiation by reacting harmlessly
with radiation-produced free radicals.

6.   Radical-molecule reactions with significant activation
energies are favoured by increasing temperatures, whereas the
rates of radical-radical reactions are generally independent of
temperature. The micro-composition of the sample depends on its

physical state (liquid or frozen) and this can affect competition between different types of radical reactions.

7. γ- and X-rays and fast electrons produce basically identical effects but the dose-rate is normally much lower with γ- and X-rays than with fast electrons, and this or other differences in irradiation conditions can result in an apparent influence of type of irradiation.

ACKNOWLEDGEMENT

Some of the work on which this article is based has been supported by grants from the Cancer Research Campaign and the Medical Research Council.

REFERENCES

1. F. Rogers, "Atomic Energy Research Establishment R4601", HMSO, London, 1964.

2. K.H. Chadwick, D.A.E. Ehlermann and W.L. McLaughlin, "Manual of Food Irradiation Dosimetry", Technical Report Series No.178, International Atomic Energy Agency, Vienna, 1977.

3. C. von Sonntag, Adv.Carbohydrate Chem.Biochem., 1980, 37, 7.

4. G.E. Adams, B.D. Michael and E.J. Land, Nature, 1966, 211, 293.

5. J. Butler, E.J. Land, W.A. Prutz and A.J. Swallow, Biochim.Biophys.Acta, 1982, 705, 150.

6. K.D. Whitburn, J.J. Shiek, R.M. Sellers, M.Z. Hoffman and I.A. Taub, J.Biol.Chem, 1982, 257, 1860.

7. A.J. Swallow, "Radiation Chemistry of Organic Compounds", Pergamon Press, Oxford, 1960.

8. A.J. Swallow, "Radiation Chemistry: An Introduction", Longman, London, 1973.

9. I.A. Taub, J.Chem.Educ., 1981, 58, 162.

10. E.J. Land and A.J. Swallow, Biochim.Biophys.Acta, 1974, 368, 86.

11. W.W. Nawar, "Radiation Chemistry of Major Food Components", Elsevier, Amsterdam, 1977, p.21.

9
The Chemistry of Nitrite in Curing

By Clifford L. Walters and Stuart I. West
LEATHERHEAD FOOD RESEARCH ASSOCIATION, RANDALLS ROAD, LEATHERHEAD,
SURREY, U.K.

INTRODUCTION

The use of nitrite in curing goes back indirectly into the realms of antiquity in that reliance was inadvertently placed upon contaminating bacteria to reduce nitrate, present in impure salt used for the preservation of meat, to nitrite. In modern times, nitrite itself is generally used alone, except in the Wiltshire curing process.

Early on the prime objective of the use of nitrite in the curing of meat was the development of the characteristic pink colour which results from the combination of myoglobin and/or haemoglobin with nitric oxide. More recently, it has been realised that nitrite and/or compounds produced from it in contact with meat contribute to the microbiological stability and freedom from the outgrowth of pathogenic organisms.

A proportion of nitrite added to meat is lost, the extent of the loss being dependent on a number of parameters. According to Nordin,[1] the rate of depletion of the nitrite concentration is related exponentially to both pH and temperature. For ham, the rate of decay was reported to double for approximately every 12°C increase in temperature and for every 0.86 units decrease of pH. Thus, the time required for the nitrite concentration in ham to be reduced to one-half of its initial value at a pH of 6.2 ranged from 513 hr at 4.4°C to only 1.6 hr at 104.4°C. When the temperature was maintained constant at 76.7°C, the half-life of the nitrite concentration was reduced from 12 hr at pH 6.7 to 4.9 hr at pH 5.6. Overall, therefore, the relationships established by Nordin were independent of the nitrite concentration over the range studied and could be expressed mathematically by the equation:

$$\log_{10} t_{\frac{1}{2}} = 0.65 - 0.025 \text{ (temperature °C)} + 0.35 \text{ (pH value)}$$

This paper is concerned with the products which can arise during the breakdown of nitrite in curing, with particular reference to the formation of the desirable haem pigments and of N-nitroso compounds.

MECHANISMS OF DEGRADATION OF NITRITE

Nitrite can be broken down by a number of processes, principally those listed below:

i) disproportionate to nitric oxide and nitrate;

ii) reduction to nitric or nitrous oxides, hydroxylamine, ammonia or nitrogen;

iii) oxidation to nitrate;

iv) reactions with phenols, amines, thiols, olefins, activated methylene groups, etc.

Additional nitrite can be regenerated from many cured meat products following treatment with mercuric or other heavy metal ions and this is often referred to as 'bound' nitrite. Ascorbate is often used during curing and this usually leads to lower nitrite levels.

GASEOUS PRODUCTS FROM NITRITE

Solutions of nitrite are reasonably stable at a pH value typical of that of meat (6.0) and disproportionation is not marked. At pH 4.0, however, breakdown to nitric oxide is evident by the formation of nitrogen dioxide above the solution.

It has been demonstrated that when nitrite is in contact with meat and incubated anaerobically at pH 6.0, both nitric and nitrous oxides are formed.[2] Part of the nitric oxide formed can be located in combination with the endogenous haem protein of the meat. Using nitrite labelled with isotopic nitrogen, Cassens et al.[3] found that usually 70-80% of the ^{15}N could be recovered from meat, of which 1-5% was released in a gaseous form. Of the remainder, 5-15% was combined with haem protein; 5-20% and 1-10% respectively remained as residual nitrite or were converted to nitrate. In addition, 5-15% of the ^{15}N was bound to thiols, 1-5% to lipids and 20-30% to proteins. Studies at the Leatherhead Food R.A., also using nitrite labelled with ^{15}N, have shown the presence of compound(s) in aqueous extracts of cooked cured meat, which contain 12% of input nitrite nitrogen and which elute from a gel permeation column coincidentally with inosine monophosphate, of approximately 400 Daltons molecular weight.

REACTIONS WITH HAEM PROTEINS

The primary action of nitrite in relation to the haem proteins myoglobin and haemoglobin is to oxidise the natural oxygenated to the ferric met-forms. The latter are brown or grey and generally distasteful in appearance to the shopper. Subsequently reduction occurs both of nitrite to nitric oxide and of the ferric to the ferrous forms of haem protein to produce the desirable pigments nitrosylmyoglobin and nitrosylhaemoglobin.

It is generally accepted that the residual respiratory enzymes still active in skeletal muscle of the slaughtered animal are at least partly responsible for the reduction of the ferric forms of the haem proteins, and probably that of nitrite to nitric oxide to complex with their ferrous forms in producing the pink colour characteristic of cured meat products. A number of possible routes exists by means of which metmyoglobin (or -haemoglobin) can be converted into nitrosylmyoglobin, as is illustrated in Fig.1. Nitric oxide can combine readily with metmyoglobin to form nitrosylmetmyoglobin (NOMetMb) and spectroscopic evidence for its formation as an intermediate has been obtained in model systems involving ascorbate. This haem derivative is unstable and is even auto-reduced on standing to

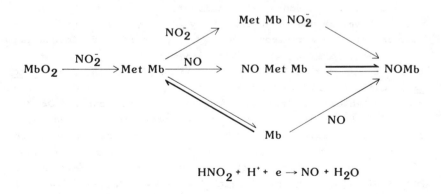

Figure 1 Possible routes of formation of nitrosylmyoglobin (NOMb) from oxymyoglobin (MbO$_2$).

form the corresponding ferrous form[4] and it is readily reduced by enzyme systems active in skeletal muscle mitochondria in the presence of reduced nicotinamide adeninedinucleotide (NADH). Alternatively, metmyoglobin itself could be reduced to deoxygenated myoglobin (Mb) before combining with nitric oxide. This reduction proceeds enzymically with great difficulty probably because the product, Mb, is rapidly reoxidised to MetMb at the low oxygen tensions typical of the interior of a block of meat. In the presence of nitric oxide, however, the ferrous form, Mb, would be stabilised by complexing with this ligand. Of the two reduction mechanisms outlined, therefore, that of nitrosylmetmyoglobin proceeds more readily than that of uncomplexed metmyoglobin except when preformed nitric oxide is present. Metmyoglobin (and methaemoglobin) form spectro-scopically defined nitrite salts and again there is evidence for their formation in model systems before the introduction of ascorbate as a reductant.[5] All of the haem pigment changes described can be brought about by ascorbate, which is widely used as a curing additive.

So far as the necessary reduction of nitrite to nitric oxide is concerned, Walters and Taylor[6] obtained evidence that nitrite could act as a terminal electron acceptor to the respiratory chain of the muscle tissue, though with much less affinity than that towards the physiological acceptor, namely oxygen. As a result the formation was observed of a complex of nitric oxide with ferricytochrome c, which on re-reduction gave rise to uncomplexed ferrocytochrome c, thereby permitting the transfer of the nitrosyl group to another receptor such as metmyoglobin. The actual enzyme involved in the reduction of nitrite, namely cytochrome oxidase or cytochrome a_3, also complexes with nitric oxide in the ferrous form and this nitrosyl derivative may also participate in the process of reduction.

Cheah[7] is of the opinion that the reduction of MetMb formed in cured meat can be accomplished by the reduced nicotinamide adeninedinucleotide (NADH), which can be formed from NAD^+ by the action of lactate dehydrogenase in the presence of lactate.

In the late 1930's, Brooks[8] showed that in the absence of oxygen equimolar quantities of nitrosyl- (NOHb) and methaemo-globin were formed from the reaction of haemoglobin with nitrite. In the presence of a reducing agent such as dithionite,

only NOHb was formed.   It has been suggested by Koizumi and
Brown[9] that a similar reaction between deoxygenated myoglobin can
reduce   nitrite   to   nitric   oxide   directly   so   as   to   form
nitrosylmyoglobin from further deoxygenated myoglobin.   However,
Lougevois[10] was unable to detect any deoxygenated myoglobin in a
system incorporating ascorbic acid.

The conversion of myoglobin to nitrosylmyoglobin is rarely
complete in an uncooked cured meat product despite a large excess
of nitrite on a molar basis.   In spite of this, the reducing
systems in meat are capable of converting added metmyoglobin to
an   overall   extent   greater   than   that   of   the   endogenous   haem
pigments present.   It has therefore been suggested by Lougevois[10]
that the overall reaction is reversible, the extent of formation
of   the   nitrosyl   pigment   being   dependent   upon   the   level   of
myoglobin available.

Residual enzyme activity would not be available in cooked
cured meat products and in these the intervention of sulphydryl
compounds is probable, as is discussed later.

REACTIONS WITH MUSCLE COMPONENTS

Undissociated nitrous acid is the reactive form of nitrite
and hence its reaction with muscle components will generally be
facilitated by pH values below its pK_a of 3.27.   Nevertheless,
low levels of undissociated nitrous acid are still available at
pH values of up to a value of about 6, at which level its
concentration approaches zero asymptotically.   The temperatures
achieved in cooked cured products will also assist in promoting
reactions   of   nitrite,   which   are   sluggish   at   ambient
temperatures.   Furthermore, nitrite can give rise to oxides of
nitrogen, some of which are very reactive on the alkaline side of
neutrality.   Some of its reactions with muscle components are
considered to be reversible whilst in others the products are
volatile and/or unreactive; these reactions are likely to proceed
to completion.

Although not a muscle component, ascorbate is used widely in
curing.   It reacts readily with nitrite, one of the products of
the reaction being proposed as being the alkyl nitrite formed at
carbon-3.[11]   Nitric oxide is also produced and therefore is
available for the formation of the cured meat pigment.   Ascorbate
is also useful in reducing the formation of N-nitrosamines in

cured products. However, the overall outcome will depend upon whether or not the nitric oxide produced from its reaction with nitrite is allowed to combine with oxygen to form the very reactive oxides N_2O_3 or N_2O_4.

AMINES

The reaction of nitrite with primary amines, such as the alpha-amino groups of amino acids, has long been featured in the Van Slyke determination of such groups in physiological samples by means of the nitrogen gas liberated. It has been suggested by Scanlan[12] that nitrous acid can form carbonium ions from primary amines, which can then interact with a molecule of unchanged amine to form a secondary amine capable of forming an N-nitroso derivative. This reaction occurs most readily in a polyamine by means of an intramolecular reaction, in which yields of up to 10% have been reported.

The reaction of nitrite with a secondary amine is considered to be reversible, the rate of formation of a N-nitrosamine being dependent upon the composition of the medium. It requires both the amine and the nitrous acid to be in the undissociated form and thus it proceeds most readily at a pH value of around 3.5. At more acidic pH values, the increasing protonation of the amine tends to suppress the reaction, whilst at higher pH values, the proportion of nitrite as undissociated nitrous acid diminishes. Of even greater importance, however, is the basicity of the amine involved. The rate of nitrosation of piperazine[13] (pK_a = 5.57) at the optimum pH, for instance, is 185,000 times that of piperidine (pK_a = 11.2). In general, the rate of nitrosation of a secondary amine is proportional to the square of the nitrite concentration and directly to that of the amine. In the case of amides, however, the rate of nitrosation is directly proportional to the nitrite concentration and continues to increase with hydrogen ion concentration throughout the pH range. Amides, in general, have much less tendency to protonate than secondary amines. Throughout most of this century, it has been taught that tertiary amines do not react with nitrous acid and this fallacy has been incorporated into a procedure for the differentiation of primary, secondary and tertiary amines. It is now apparent that many tertiary amines do, in fact, react with nitrous acid to form the N-nitrosamine(s) of the secondary amine(s) resulting from the

cleavage of an alkyl, aryl or heterocyclic group together with a
carbonyl compound. This so-called 'nitrosative cleavage' is
applicable to many types of tertiary amines; from nicotine, for
instance, a methyl group is cleaved during reaction with nitrous
acid to form N-nitrosonornicotine. The formation of N-nitroso-
dimethylamine (NDMA) from the analgesic aminopyrine occurs much
more readily on reaction with nitrous acid or oxides of nitrogen
than that from its parent amine, dimethylamine. Quaternary
ammonium compounds have been reported to break down similarly.
So far as meat is concerned, this could have particular
importance in relation to its contents of carnitine and choline.
The formation of N-nitrosopyrrolidine and a number of other
N-nitrosamines from spermidine has been reported,[12] whilst
creatinine can react with nitrous acid to form N-nitroso-
sarcosine; this leads to NDMA on decarboxylation during the
cooking process.

It was found by Keefer and Roller[14] that the pH range over
which the nitrosation of a secondary amine occurs extends well
into the alkaline region in the presence of formaldehyde or other
simple aldehydes. Probably of even greater importance in this
context, however, are the nitrosating capacities of the oxides of
nitrogen N_2O_3 and N_2O_4, which increase with pH value, according
to Challis et al.,[15] far beyond the range typical of meat even
when an alkaline phosphate is used as an additive. As mentioned
previously, ascorbate can inhibit the process of nitrosation. In
model system studies,[16] it has been shown to inhibit strongly the
formation of the N-nitroso derivatives from a range of amines and
amides, but not from dimethylamine. However, this is a very
basic amine and its extent of nitrosation was very small under
the conditions employed. The formation of NDMA in Frankfurters
prepared with excess nitrite,[17] was found to be markedly reduced
when ascorbate was used, but not always eliminated, even when the
ascorbate level was increased ten-fold to 5,500 ppm. A number of
other antioxidants and/or nitrite scavengers, such as
alpha-tocopherol, have also been reported to be effective in
reducing the formation of volatile N-nitrosamines in cured meat
products. Sen et al.[18] and Walters et al.,[19] for instance, found
ascorbyl palmitate to be more consistent than ascorbate itself in
inhibiting the formation of volatile N-nitrosamines in fried
bacon. This could result from the finding by Mottram et al.[20]

and by Coleman[21] that N-nitrosopyrrolidine (NPyr) and NDMA are
formed mainly in the adipose tissue during the frying of bacon
and ascorbyl palmitate is preferentially lipid soluble. It was
postulated by Bharucha et al.[22] that the site of formation of
NPyr resulted from the higher temperatures achieved by the
adipose tissue in comparison with those in the lean, but Mottram
et al.[20] concluded that the non-polar lipid nature of the adipose
tissue provided an environment conducive to the formation of
N-nitrosamines. However, Massey, et al.[23] observed a 20-fold
enhancement in the formation of N-nitrosodihexylamine in the
presence of decane in a model system but no effect on that of
N-nitrosopyrrolidine. Under this condition, ascorbic acid
enhanced the nitrosation of lipophilic amines when decane was
present but did not affect the yield of N-nitrosopyrrolidine. An
alternative explanation for the localisation of NOPyr in the
lipid phase, however, is the intermediate formation of
pseudonitrosites of unsaturated lipids, such as that of
palmitodiolein, which is an active nitrosating agent under frying
conditions.[24]

The potential importance of N-nitroso compounds is that the
vast majority of the 200+ such compounds tested are carcinogenic
in experimental animals, often at very low dosages.
N-nitrosamines are of two distinct types - those volatile in
steam, which can thereby be readily separated from a food matrix,
and those which, along with the N-nitrosamides, are too complex
and/or unstable to be selectively removed in this way. The most
consistent source of volatile N-nitrosamines amongst cured meats
is fried bacon, the principal contaminants being NPyr and NDMA,
which can occur to extents within the ranges 0-50 and 0-20 µg/kg,
respectively, but generally towards the lower limits,
particularly after the addition of ascorbate. During the process
of frying, considerable proportions of NPyr and NDMA formed are
distilled off and thereby lost from the bacon itself.[25] Recently
N-nitrosothiazolidine has also been detected in some cured meat
products and it is thought to be derived from cysteine and
formaldehyde.

In order to determine the contents of NPyr and NDMA in bacon
or other meat products, use is generally made of gas
chromatography coupled with a specific detector. That most
commonly employed is the Thermal Energy Analyser (TEA), in which

each volatile N-nitrosamine is cleaved catalytically in turn to nitric oxide for detection using a chemiluminescence analyser. Thus the nitric oxide liberated is oxidised using ozone to activated nitrogen dioxide, which, as it decays to its ground state, emits light in the far visible and near infra-red regions, from which it can be determined with sensitivity and selectivity. Although a response to the TEA is given by other compounds containing oxygenated nitrogen atoms, it has proved to be remarkably specific for the determination of volatile N-nitrosamines. However, it is undoubtedly true that the great majority of precursors in a biological system would give rise to non-volatile N-nitrosamines and N-nitrosamides. A method has been devised by Walters et al.[26] for the selective determination of N-nitroso compounds as a group, which involves the use of a chemiluminescence analyser designed for the measurement of nitric oxide. Initially, nitric oxide is released from any heat labile compounds such as S-nitroso derivatives of thiols or pseudonitrosites of unsaturated lipids. The addition of acetic acid leads to the further evolution of nitric oxide from any nitrite present. When gas evolution has subsided, hydrogen bromide brings about the denitrosation of N-nitroso compounds, again to nitric oxide, from which they can be monitored, although similar responses can be given also by S-nitrothiols and nitrolic acids. Finally, nitric oxide can be derived from nitrate and C- and N-nitro compounds by the subsequent introduction of titanous chloride. This method has proved useful in detecting non-volatile nitrite derived compounds from cured meat. At the Leatherhead Food R.A., for instance, this system has been used in the detection of compounds in aqueous extracts of cured meat separated by gel permeation chromatography.

PHENOLS

The smoking of cured meats has been stated[27] to deposit up to 300 ppm of phenols on the matrix, so that this type of contaminant can be of great importance in relation to nitrite breakdown. The initial product of the nitrosation of a phenol is a C-nitrosophenol but many of these are unstable under aerobic conditions and oxidise to form the corresponding C-nitrophenol. According to Challis,[28] the reaction of nitrite with phenols proceeds more rapidly over a wide range of pH on the acidic side

of neutrality than does the nitrosation of a basic secondary
amine such as dimethylamine. As scavengers of nitrite,
therefore, phenols would be expected to act as inhibitors of the
formation of N-nitroso compounds in a complex biological system
in which both were present. The outcome, however, depends upon
the structure of the phenol(s) concerned, according to Davies et
al.,[27] who found that C-nitroso phenols can catalyse the
N-nitrosation of secondary amines. They studied the structural
requirements for the catalytic effect and deduced that all of the
compounds they found to stimulate the nitrosation of a secondary
amine were capable of tautomerisation to form a quinonemonoxime
or a quinonemonoxime imine. Thus, p-nitrosophenol acts as a
catalyst to the nitrosation process but not when the hydroxyl
group is blocked by tosylation. Other catalytic C-nitrosophenols
include p-nitroso-cresol and l-nitroso-2-naphthol. It was also
found that p-nitroso-N-alkylanilines and p-nitroso-N,
N-dialkylanilines can catalyse the nitrosation of secondary
amines. The suggested mechanism of catalysis by C-nitrosophenols, p-nitroso-N-alkylanilines and p-nitroso-N,N-dialkyl-
anilines involves the initial nitrosation of the catalysts either
in or via the quinone monoxime form to yield an intermediate
which is then attacked by the amine to form the N-nitrosamine and
to regenerate the catalyst. Using nitrite labelled with ^{15}N,
Pignatelli[29] was able to show that the N-nitroso group of NDEA
formed by catalysis with p-nitrosophenol was derived from nitrous
acid and not by transnitrosation from the C-nitroso compound.

Initially, therefore, the rate of nitrosation of a secondary
amine will be reduced in the presence of a 'catalytic' phenol
unless the nitrite concentration is so great as to be essentially
constant. As nitrosation of the phenol takes place, catalysis of
the nitrosation of the amine will be increasingly evident.
According to Davies et al.,[27] the overall effect will depend upon
whether the steady state concentration of the C-nitrosophenol, as
determined by the relative rates of its formation and of its
oxidation to the C-nitrophenol, is sufficiently large to overcome
the inhibitory effect of the phenol itself.

THIOLS

Thiols also react more rapidly with nitrite at acidic pH
values than dialkylamines, the main products being their

S-nitroso derivatives. These are generally unstable compounds which are considered to contribute to the so-called 'bound' nitrite which can be released from cured meat products on treatment with a mercuric or other heavy metal salt.[30] Mirna and Hofmann[30] also suggested that S-nitrosothiols could act as nitrosating agents in cured meat products. This suggestion was investigated by Davies et al.[27] who compared the relative efficiencies of an equimolar level of S-nitrosocysteine and of nitrite itself in nitrosating N-methylaniline at various pH values and at 37°C. At pH 2.65, nitrite was a much better nitrosating agent than S-nitrosocysteine but this difference had almost vanished on raising the pH to 5.5. Transnitrosation from S-nitrosocysteine to N-methylaniline continued, albeit at a reduced rate, when the pH was increased further to 9.75, at which value no formation at all of the N-nitrosamine occurred using nitrite. However, the main route of decomposition of an S-nitrosothiol involves the formation of the disulphide with liberation of nitric oxide. Thus, it has been found that S-nitrosothiols are active in the conversion of metmyoglobin into nitrosylmyoglobin, particularly in the presence of a reducing agent such as cysteine itself or ascorbate. The reducing systems available in meat are also able to potentiate the conversion by S-nitrosocysteine of met- to nitrosylmyoglobin.

As well as their capacities to form the nitrosyl pigment, it has been suggested that S-nitrosothiols formed during curing contribute to the stability and safety of such products by inhibiting bacterial growth. In practice, moreover, the amounts of volatile N-nitrosamines formed during the frying of bacon prepared using an S-nitrosothiol were generally markedly less than those derived from similar samples prepared with equimolar levels of nitrite. Furthermore, similar stabilities have been achieved in the two products in spite of the much lower levels of residual nitrite in those produced with S-nitrosocysteine.

OLEFINS

Nitrous acid and its anhydride N_2O_3 can react with unsaturated compounds, such as lipids, with the formation of a pseudonitrosite. The potential importance of the formation of such compounds in the context of curing is that they are capable of acting as nitrosating agents in model systems at a temperature

(170°C) typical of the frying of bacon. The pseudonitrosite of the lipid palmitodiolein, for instance, can bring about the formation of N-nitrosomorpholine from its parent amine.[24] This effect may contribute to the localisation of the volatile N-nitrosamine NPyr in the adipose tissue after frying.

ACTIVATED METHYLENE GROUPS

According to a review of the subject,[31] the reaction of nitrous acid with methylene groups activated by adjacent carbonyl, carboxyl or other groups leads generally to subsequent rearrangement to form oximes, except where tertiary carbon atoms are involved. Thus, C-nitro derivatives at a primary carbon atom give rise to nitrolic acids, an example of which has been reported in corn treated with nitrous acid, along with nitro-hexane.[32] Similar derivatives at a secondary carbon atom yield pseudonitroles on treatment with nitrous acid.

CONCLUSIONS

From a microbiological point of view, meat products prepared using nitrite have a good record of safety and stability; this results, in part at least, from the residue of nitrite and/or products derived from its contact with muscle components. In appearance meat products so treated are generally attractive to the purchaser by virtue of the haem pigments stabilised through combination with nitric oxide. Whether or not nitrite or products derived from it contribute to flavour is still a matter of conjecture. There is little doubt, however, that the use of nitrite can lead to the formation of trace amounts of volatile N-nitrosamines and probably to other compounds following reactions with phenols, thiols, olefins, etc. In evaluating the potential risks to the consumer, however, consideration should also be given to his or her exposure to other sources of nitrate, notably in vegetables and water supplies, which can give rise to relatively high levels of nitrite *in vivo*, and to the probable biosynthesis of nitrate.

REFERENCES

1. H.R. Nordin, J.Inst.Technol.Aliment., 1969, 2 (2), 79.
2. C.L. Walters and R.J. Casselden, Z.Lebensm.Unters.Forsch., 1973, 15, 335.

3. R.G. Cassens, G. Woolford, S.H. Lee and R. Goutefongea, "Proc.2nd Int.Symp.Nitrite Meat Prod.", Pudoc, Wageningen, 1977, p.95.

4. D. Keilin and E.F. Hartree, Nature, Lond., 1937, 139, 548.

5. J.B. Fox and J.S. Thomson, Biochem., 1963, 2 (1), 465.

6. C.L. Walters and A.McM. Taylor, Biochim.Biophys.Acta, 1965, 96, 522.

7. K.S. Cheah, J.Food Technol., 1976, 11, 181.

8. J. Brooks, Proc.Roy.Soc.Lond., 1937, B123, 368.

9. C. Koizumi and W.D. Brown, J.Food Sci., 1971, 36, 1105.

10. V.P. Lougevois, "The Formation of Cured Meat Colour", Ph.D. Thesis, University of Leeds, 1982.

11. H. Dahn, L. Loewe and C.A. Bunton, Helv.Chim.Acta., 1960, 43, 320.

12. R.A. Scanlan, CRC Crit.Rev.Food Technol., 1975, 5, 357.

13. S.S. Mirvish, "IARC Scientific Publications No.3", IARC, Lyon, France, 1972, p.104.

14. L.K. Keefer and P.P. Roller, Science, 1973, 181, 1245.

15. B.C. Challis, A. Edwards, R.R. Hunma, S.A. Kyrtopoulos and J.R. Outram, "IARC Scientific Publications No.19", IARC, Lyon, France, 1978, p.127.

16. S.S. Mirvish, L. Wallcave, M. Eagen and P. Shubik, Science, 1972, 177, 65.

17. W. Fiddler, J.W. Pensabene, E.G. Piotrowski, R.C. Doerr and A.E. Wasserman, J.Food Sci., 1973, 38, 1084.

18. N.P. Sen, B. Donaldson, S. Seaman, J.R. Iyengar and W.F. Miles, J.Agric.Food Chem., 1976, 24, 397.

19. C.L. Walters, M.W. Edwards, T.S. Elsey and M. Martin, Z.Lebensm.Unters.Forsch., 1976, 162, 377.

20. D.S. Mottram, R.L.S. Patterson, R.A. Edwards and T.A. Gough, J.Sci.Food Agric., 1977, 28, 1025.

21. M. Coleman, J.Food Technol., 1978, 13, 55.

22. K.R. Bharucha, C.K. Cross and L.J. Rubin, J.Agric.Food Chem., 1979, 27, 63.

23. R.C. Massey, M.J. Dennis, C. Crews, D.J. McWeeny and R. Davies, "IARC Scientific Publications No.32", IARC, Lyon, France, 1980, p.291.

24. C.L. Walters, R.J. Hart and S. Perse, Z.Lebensm.Unters. Forsch., 1979, 168, 177.

25. T.A. Gough, K. Goodhead and C.L. Walters, J.Sci.Food Agric.,

1976, <u>27</u>, 181.

26. C.L. Walters, M.J. Downes, M.W. Edwards and P.L.R. Smith, <u>Analyst</u>, 1978, <u>103</u>, 1127.

27. R. Davies, M.J. Dennis, R.C. Massey and D.J. McWeeny, "IARC Scientific Publications No.19", IARC, Lyon, France, 1978, p.183.

28. B.C. Challis, <u>Nature, Lond</u>., 1973, <u>244</u>, 466.

29. B. Pignatelli, "Etude de l'influence des Produits Phenoliques sur la Vitesse de Formation d'un cancerogene chimique: la N-Nitrosodiethylamine," Thesis, Universite Claude-Bernard, Lyon, France, 1979.

30. A. Mirna and K. Hofmann, <u>Die Fleischwirtschaft</u>, 1969, <u>49</u>, 1361.

31. P.A.S. Smith and D.R. Baer, <u>Org.React</u>., 1960, <u>11</u>, 157.

32. M.C. Archer, T.J. Hansen and S.R. Tannenbaum, "IARC Scientific Publications No.31", IARC, Lyon, France, 1980, p.305.

10
Some Developments in Analytical Techniques Relevant to the Meat Industry

By Ronald L. S. Patterson

A.R.C. MEAT RESEARCH INSTITUTE, LANGFORD, BRISTOL BS18 7DY, U.K.

INTRODUCTION

Many analytical techniques can be applied to meat and meat products; precisely which are used depends upon the purpose of an investigation, the time and money available, and the amount of detail required in the information. The analytical chemist can view meat simply as a composition of lean, fat, skin and bone, differing only in relative proportions from piece to piece; or he or she may be interested in the more detailed structure of the lean or fat, for example, the amount of connective tissue or intramuscular marbling fat present, or the exact fatty acid composition of the adipose tissue.

The question of inorganic additives (curing salts) or contaminants (heavy metals) may be of particular importance, or residues of organic compounds such as herbicides and pesticides or other substances which accumulate during the lifetime of the animal. Of much interest at the present time is the possible presence of residues of anabolic agents, the hormonal substances introduced deliberately into meat animals to enhance growth. After slaughter, contamination of carcass meat can still occur during the storage and distribution stages through misuse or accidental spillage of chemicals used as disinfectants or detergents, causing organoleptic problems which often become manifest only during cooking or eating.

The range of techniques available is wide, encompassing mass spectrometry, nuclear magnetic resonance and infrared spectrometry, gas-liquid chromatography, atomic absorption spectroscopy, etc., as well as modern versions of traditional wet chemical procedures. This paper does not attempt to discuss every development which has taken place in analytical chemistry in the last few years, but rather to select some areas where developments have been particularly interesting or are new in concept, and are relevant to the meat industry.

MEAT SPECIATION

A question which arises sometimes is whether or not meat is actually from the declared species of animal. Although improvements in technology and processing have led to more economical transportation and utilisation of deboned carcass meat, this has also facilitated the use of cheaper undeclared meats, because unequivocal identification of species becomes very difficult once meat has been taken off the carcass and the anatomical features have been destroyed. Meat of similar pigmentation, for example beef and horse meat, beef and mutton, or poultry and pig meat, are virtually impossible to distinguish by eye once they have been frozen en masse in large blocks, or flaked and incorporated into comminuted meat products.

However, apart from deliberate misrepresentation, many meat products from continental Europe as well as British products, such as sausages, burgers and pies, may contain legally the flesh of more than one species, and the analyst requires methods of species identification; with current legislation a quantitative estimation of any species flesh content is of practical importance.

Methods for the determination of species origin have been available for some time based upon immunological antigen/antibody reactions using various forms of the precipitin test. Precipitating antisera for the different meat species are available commercially and can be used qualitatively in Ochterlony-type double immunodiffusion tests[1] or semi-quantitatively in immunoelectrophoresis.[2] One disadvantage of these methods is that they require concentrated antibody preparations which becomes very expensive in large scale testing. Also the antisera may not be 'pure', having been raised against mixed antigens, and this can result in more than one line or ring of identity, leading to uncertain results. Even when the test species are phylogenically remote, some non-specific cross-reaction(s) may occur, making these methods ambiguous. Gel electrophoresis[3] and isoelectric focussing[4,5] are alternative techniques which have had considerable success in identification of the species origin of fresh meats and fish, but which are unsuitable for the analysis of mixtures containing more than one species.

Enzyme-linked assays

Enzyme-linked immunosorbent assay (ELISA)[6],[7] has emerged in the last few years as a rapid, convenient and relatively cheap method of assaying antigens and antibodies quantitatively in many diagnostic tests in clinical medicine. Recently, we adapted a particular form of ELISA in meat species identification.[8] The technique has also been applied to the detection and estimation of soya protein in food products,[9] and it is clear that many more applications will be developed in future for identification of food components capable of acting as antigens.

Many antisera are now commercially available, and although produced in a host animal (rabbit, sheep, goat) in response to injection of a **single** antigenic substance, for example, a serum albumin, they comprise a number of different types of antibodies depending upon the number of antigenic determinants (sites) present on the injected antigen. Such antisera are termed heterologous, **polyclonal** antisera. Cross-reacting antibodies can be removed in many cases by immunoadsorbent chromatography[10] but where a very close phylogenic relationship exists between species, e.g. sheep and goats, then affinity chromatography may be ineffective. Nevertheless, after an antiserum has been purified, rendered as monospecific as possible and standardised for use, it may serve the required purpose quite satisfactorily for as long as the supply lasts, and provided no other unexpected crossreacting components appear. However, when exhausted, a fresh batch will be required, and this immediately reveals another major disadvantage of conventional antisera, that is the lack of reproducibility from batch to batch. A second immunisation is quite likely to produce a new antiserum with different properties from the first because the component antibodies have been produced in different proportions, and so the purification and standardisation procedures must be repeated.

At the present time, **monoclonal** antibodies appear to be a panacea for (many of) the faults of polyclonal, heterologous antisera. A monoclonal antiserum is composed of only one type of antibody, specific to only one of the many possible antigenic determinants of the parent antigen, and so is 100% pure. The great advantage of monoclonal antisera is that once the desired characteristics for an antiserum have been selected, production can be initiated in a controlled system which is potentially everlasting. Methods of production of monoclonal antibodies differ considerably from those methods for conventional antisera. They are based upon cell-fusion techniques in which myeloma (cancer) cells and normal spleen cells are fused in vitro to form new hybrid cells. The new cells retain some aspects of cancer, i.e. vigorous growth, and some aspects of normality, e.g. antibody production. Although very many clones may have to be evaluated initially, once one has been found with precisely the desired characteristics of specificity required for a particular assay, then culture of that clone can continue in vitro or in the abdominal cavity of a suitable host animal (mouse) as a tumour. A solution of monoclonal antibodies is therefore the product of a single clone of antibody-secreting cells that have the capacity for endless reproduction.

Species identification - by ELISA

Monoclonal antibodies are not yet available commercially for identification of the common meat species, so conventionally-produced antisera were used in our adaption of the ELISA test.[8] The antisera (purified by affinity chromatography) had all been raised in rabbits and were therefore rabbit immunoglobulins. The enzyme used to visualise and quantify all the assays was horse radish peroxidase conjugated to an antibody to rabbit immunoglobulin, i.e. an anti-antibody. In the presence of substrate, increased development of the yellow chromophore, measured as absorbance at 492 nm, indicated increased antigen content in the original meat extract. Horsemeat was clearly differentiated from beef, pig and mutton at the optimum dilution of the meat extracts (Fig.1a). It was also detected easily in mixtures with beef at levels between 3 and 80% (Fig.1b); although small differences in absorbance values were found for beef containing less than 3% horsemeat compared with those for pure beef, they were not statistically significant. Fig.1c shows clear differentiation of beef and veal from pig, horse and mutton with an anti-BSA serum, whilst the cross-reactivity of the anti-SSA serum with goat serum albumin is seen in Fig.1d, due to the very close phylogenic relationship of the two species: a degree of cross-reactivity of this antiserum with beef serum albumin is also evident.

These results demonstrate that a conventional serum may not

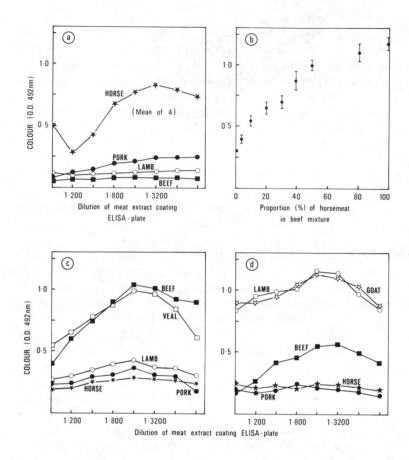

Figure 1 Identification of meat species by ELISA: (a) clear differentiation of horsemeat from pork, lamb and beef; (b) increasing colour development as horsemeat content of beef mince increases from 3%; (c) differentiation of beef and veal from pork, horse and lamb by an antiserum to bovine serum albumin; (d) cross-reaction of sheep and goat meats (strong) and beef (weak) with an anti-sheep serum. Note that the dilution factor for the meat extract coating the ELISA plate must be optimised to achieve maximum differentiation.

be fully mono-specific even after purification by affinity chromatography and that some care should be exercised when interpreting results obtained from sera purified in this way. It should be noted that the use of antisera raised against serum albumins or other native proteins will be effective only in the differentiation of raw, unheated meats: a completely new and different set of antisera specific to heat-stable proteins will be required for heat processed and cooked meats.

Species identification - by direct probe mass spectrometry

A completely different and novel approach to identification of the species origin of a meat sample is by the use of direct probe mass spectrometry (DPMS). The technique has been explored in this laboratory as a possible rapid means of partial characterisation and differentiation of micro-organisms.[11],[12] DPMS is a lower temperature version (less than 300°C) of pyrolysis-mass spectrometry[13],[14] which generally involves temperatures between 500 and 800°C.

For analysis by this technique, samples of meat are prepared in the form of freeze-dried lyophylates. Approximately 50 μg is introduced by platinum wire into a quartz glass pyrolysis tube held in the end of the direct probe. Specimens are heated in the

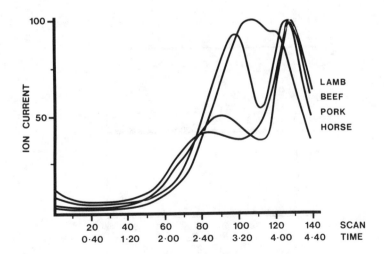

Figure 2a Ion current profiles of meat samples

ion source of the mass spectrometer from ambient to 300°C at 60°C per min.   Mass spectra are recorded repeatedly over the mass range m/e 33 to 400 using a 2 sec scan cycle. As the temperature of the probe increases the total ionisation produced from the sample is recorded as a function of time in an ion current profile. A single spectrum is then produced for each sample by averaging across the profile and subtracting a ten-scan background to remove any underlying contamination.

Typical ion current profiles are shown in Fig.2a: those for horse and lamb are quite different and distinct from those of pork and beef which closely resemble one another.   However, the shape of the profile does not always reflect the origin of the sample and the averaged spectra are a much more reliable guide (Fig.2b). As is often the case with DPMS of biological material,

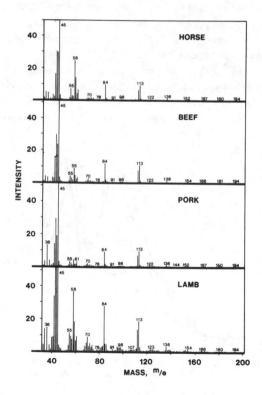

Figure 2b Averaged spectra of meat samples

the spectra have strong qualitative similarities and differ only in the relative intensities of the individual ions: this implies that the potential of the method for analysing mixtures of different meats is probably not very good.   Spectra are then converted to a Fortran readable format, normalised to remove variations in sample concentration and analysed by multivariate data analysis routines.   The data are displayed either as a simple scatter diagram, where the normalised intensities of the two most characteristic ions are plotted against each other, or as a non-linear map.   The latter, which is a two-dimensional representation of the original multidimensional similarity relationships between the samples, is produced by multi-dimensional scaling of the matrix of similarities between the spectra, calculated using a subset of up to 20 of the most characteristic ions.

In the data illustrated, the statistical procedures showed that the ions m/e 38 and 36 provided the greatest degree of discrimination, followed by m/e 74 and 58.   Since m/e 38 and 36 were shown to be highly correlated, a plot of m/e 36 against m/e 74 was prepared (Fig.3a) which shows clear separation of the

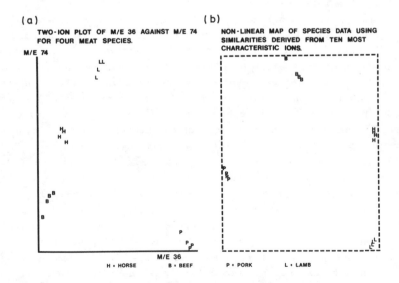

Figure 3   Discrimination of meat species by DPMS

species.   Multidimensional scaling using the 10 most characteristic ions   is shown in Fig.3b.   Both data analysis procedures
show that the four meats can be clearly distinguished from one
another, with the replicates clustering well together.   The same
techniques were also used in two, more comprehensive studies.   In
the first, six **different muscles** were examined from each of three
animal species, beef, chicken and rabbit.   Clear separation of
the species was achieved: Fig.4a is a plot of m/e 70 against m/e
61 and Fig.4b is a non-linear map obtained by using the 20 most
characteristic ions in the calculation of similarities between
samples.   Both procedures are equally effective in resolving the
data.   In the second study, one sample was taken from each of
four individual animals from each of six species, beef, horse,
pig, sheep, chicken and rabbit.   These data are shown in Fig.5 as
a non-linear map derived using the three most characteristic ions
from this data set.   Discrimination into groups is still
reasonably clear even when six species are analysed in the same
study.   Although still in its infancy, the technique appears to
have considerable potential awaiting development.

PROTEIN ANALYSIS
     Information about the protein content of a meat product is
frequently required.   The classical method of protein analysis is
by Kjeldahl,[15] in which the total nitrogen is estimated.
Conversion of the weight of nitrogen found to a percentage basis
for the wet, fat-free lean gives the nitrogen factor for the
sample which, when multiplied by 6.25 (for beef, on the
assumption that the protein contains 16.0% nitrogen), gives a
measure of the percentage of crude protein in the sample.
Average N-factors and protein conversion factors are: beef 3.55
and 6.25; pork 3.45 and 6.20; and lamb 3.2 (approx.) and 6.00.
However, all nitrogen-containing substances present whether
proteinaceous or not are assayed by the Kjeldahl procedure, so
results for 'protein' obtained in this way can be high (up to
12%).   One way of avoiding this error is to carry out duplicate
nitrogen analyses, having first precipitated all protein from one
aliquot of the initial sample slurry; the figure for nitrogen
obtained for the protein-free solution is then deducted from that
for the other sample.   If the non-protein nitrogen figure is
significantly above 10%, then consideration must be given to

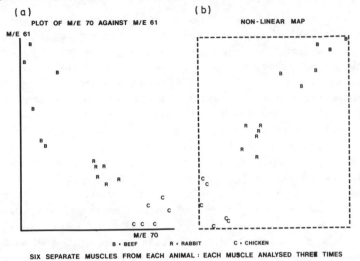

Figure 4 Discrimination of meat species by DPMS; (a) plot of m/e 70 against m/e 61; (b) non-linear map.

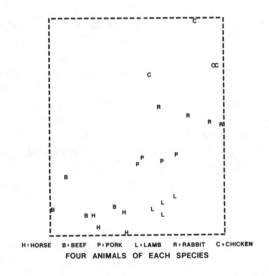

Figure 5 Discrimination of six meat species by DPMS.

possible sources of high levels of non-protein nitrogen and examination made for substances such as urea or ammonium salts.

In meat products, the protein may be of animal or vegetable origin. A number of methods of estimating the cereal, milk, soya and other non-meat protein constituents have been evolved,[16] and procedures to account for all the possibilities are complex and protracted. Estimation of the meat content of a product by direct measurement of a characteristic component specific to meat protein represents an alternative approach to the customary method of estimation by application of correction factors to total nitrogen analysis. Recent work by Jones et al.[17] has made considerable progress in this direction. Their method is based upon the earlier work of Lawrie and co-workers[18] who observed that the amino acid 3-methylhistidine (3-MeH) occurs at almost constant levels in the actin and myosin of skeletal muscle of the main meat species whilst being virtually absent from non-meat food proteins, such as soya; also it is stable to processing and hydrolysis.[19] Using this information, Jones's group evolved a procedure for the estimation of the meat protein content of a product based upon the fluorometric measurement of the fluorescamine derivative of 3-MeH. The method is straightforward and specific, with the exception of pork meat which requires slightly different treatment during preparation to remove interfering dipeptides. Samples are subjected to conventional hydrolysis in 6M hydrochloric acid and an aliquot neutralised prior to derivatisation with fluorescamine in borate buffer. Incubation for an hour in acid conditions results in destruction of the fluorescamine derivatives of all the amino acids except those of histidine and 3-MeH which are acid-stable. Separation of the latter pair is achieved by HPLC on a reversed-phase column, with detection by a fluoromonitor. Calibration graphs are prepared using standard solutions of derivatised 3-MeH, and results are expressed as milligrams 3-MeH per g non-collagen nitrogen. The limit of detection is about 10 ng ml^{-1} in a hydrolysate. The levels of 3-MeH found in meat from five common species (beef, pork, lamb, rabbit and chicken) were all about 5 mg g^{-1} non-collagen nitrogen, which agreed well with figures reported previously (4.8-6.4 mg g^{-1}); figures for chicken were lower than frequently obtained, with no values in excess of 6.0 being found.

When calculating the meat content of a product by this procedure, due allowance must be made for the connective tissue and fat content.[20] Non-collagen nitrogen is the difference between total Kjeldahl nitrogen and the nitrogen derived from connective tissue as determined by hydroxyproline estimation plus non-protein nitrogen, and equates to the contractile protein content of the meat. Whilst qualitatively specific, the authors point out that use as a fully quantitative method still requires a degree of caution, since 3-MeH is not present in all muscles (e.g. cardiac myosin); generally, it originates in meat from the actin component, but also from certain fast twitch muscles, thus quantitative levels in meat depend to some extent on the ratio of the fibre types present.

INSTRUMENTAL METHODS

Infrared spectrometry

Application of infrared techniques to the analysis of meat products has aroused considerable interest in the last year or two. Analysis of multi-component meat systems has always been difficult because of interaction between components and the need for separation stages; also, all the conventional standard wet chemical procedures are too slow to serve as a practical means of monitoring or controlling the various stages of meat product manufacture on a modern high-speed line. Apart from the cost and the slowness of traditional procedures, producing small samples (1-5 g) which are truly representative of bulk quantities of raw materials is extremely difficult, and therefore methods which work with larger samples should be somewhat more accurate; also, if speed of analysis is rapid, many more samples can be analysed in a given time and a more accurate assessment of composition obtained.

Two instruments utilising different parts of the infrared spectral region have been evolved for measurement of individual components of multi-component meat mixtures without the need for prior separation. Instruments such as the InfraAlyser 400 work in the Near Infrared (NIR) region, 0.8-2.5 microns (μ), whilst the Super-Scan works in the range 2.5-10 μ. The principle of operation of the InfraAlyser type is based upon **reflection** from a sample; component analysis depends upon measurement of the

spectrally diffuse reflectance from the immediate sub-surface
layers (down to 3 mm deep) of an emulsified sample: reflections
from the actual surface are ignored. Preparation of samples
requires chopping into 2-3 cm cubes prior to high-speed
blending. The emulsified sample is then smoothed into an open
cup and placed in the instrument. Once calibrated, analysis of
multi-component samples, e.g. sausages, processed meats, etc.,
for fat, protein and moisture can be carried out in less than
5 minutes each. Calibration of the instrument requires
independent analysis of replicate samples separately for fat,
protein, and moisture by conventional chemical methods and the
use of computerised statistical techniques to relate these data
to the reflectance spectra recorded for the same samples at 19
different wavelengths. Different types of product require
different sets of calibration constants but, once established,
routine analysis of a product line becomes simple and rapid.

Recently, scanning instruments (e.g. Neotec 6350) have
appeared which scan continuously across selected regions of the
NIR, e.g. 0.6-1.2 or 1.1-2.5 μ, at intervals of 1 or 2 x 10^{-3} μ.
Spectral data from 50 scans can be collected and integrated in
less than one minute. At the moment, these new instruments are
probably best considered as research tools, providing useful
spectral information which can be incorporated in the calibration
of the fixed filter instruments to improve their accuracy for
routine product analysis.

Whilst the technique is well established for the analysis of
protein, fat and moisture in grain and cereal products,[21],[22] it
is just emerging in the area of meat product analysis; according
to those who have studied it,[23],[24] several factors such as pH,
temperature, the physical state of the fat and protein, and the
'lightness' (whiteness) of samples may affect the results and
require elucidation before the technique will be able to replace
conventional analytical methods or compete with other rapid
methods.

One such rapid method is that represented by the Super-Scan
instrument which operates in the 2.5-10 μ range of the infrared.
It differs from the previous technique in that it is the
attenuation in the **transmitted** energy which is measured, and
absorbances at specific wavelengths are related to the presence
of particular chemical bonds in fat, protein and carbohydrate.[25]

The system consists of a grinding and homogenisation unit, an electronic balance for accurate weighing of the sample (ca. 11 g), and a reactor vessel in which the sample is vigorously mixed with a special alkaline dispersing reagent which solubilises the protein and emulsifies the fat. The homogeneous milk-like liquid is transferred to the cuvette of the measuring unit in which the transmission/absorption measurements are made. The data are then passed on-line to the dedicated mini-computer which contains a variety of calibration constants to cover most product types. After correction with the appropriate constants the fat, protein, carbohydrate and water percentages are produced both visually and as print-out. Like the reflectance method, the instrument must be calibrated against standard chemical methods and adjustments made to the calibration constants until minimum deviation is obtained. According to the manufacturers, different calibrations are not necessary for each manufactured product as experience has shown that products can be grouped into broad types, for example, raw meats, sausages (British), burgers, etc.[26]

The data in Table 1 were obtained in our laboratories. After initial sets of analyses had been carried out to ascertain what adjustments were necessary to the calibration constants to match our particular handling of the wet chemical techniques, new samples were analysed by both procedures. Correlation coefficients of 0.995 or greater were obtained for each of the Super-Scan vs. routine method comparisons and also low pooled standard deviations. Whilst in this set of data there was no statistically significant difference between the Super-Scan and the Büchi Kjeldahl methods for protein, there was evidence (t-test, $P<0.01$) that the Super-Scan gave higher values than Soxhlet for fat, and lower values than vacuum oven for moisture; the differences were, however, mostly within acceptable limits.

An important addition to the Super-Scan's range of analyses will be the introduction of a rapid method for estimation of collagen capable of processing 8-10 samples per hr. If successful, this development will represent a major advance in analytical techniques as it will dispense with the requirement for overnight acid hydrolysis.

Table 1 Determination of fat, protein and moisture in meat. Super-Scan IR Analyser vs. Standard methods.

| Sample | % in meat | | | | | |
| | Fat | | Protein | | Moisture | |
	S-S	Soxhlet	S-S	Büchi Kj	S-S	Vac.Oven
Neck	3.40	3.40	19.85	19.57	75.18	75.63
Shin	4.25	4.07	21.05	21.15	73.15	73.57
"	6.03	5.57	20.37	20.45	72.07	72.50
"	6.10	5.80	20.30	20.20	71.97	72.37
Neck	7.18	7.35	20.07	20.03	71.15	71.53
Shin	8.72	7.87	20.12	20.17	69.57	70.63
Neck	10.27	9.93	18.92	18.87	69.23	70.03
Brisket	12.65	12.33	18.35	18.27	67.43	67.60
Flank	16.87	16.73	18.43	18.67	63.15	63.63
"	20.72	20.53	16.60	17.23	61.10	60.83
"	25.93	25.57	15.78	16.30	56.72	57.03
"	27.27	26.93	17.18	17.43	53.98	54.63
"	27.97	27.57	16.83	16.93	53.63	54.33
"	30.03	30.40	15.08	15.53	53.25	53.03
Correlation Coefficient	0.999		0.995		0.999	
Pooled Standard Deviation	0.131	0.125	0.136	0.194	0.187	0.107
Mean difference	0.239		-0.134		-0.411	
S.E. differences	0.078		0.069		0.095	
Paired t value	3.06**(13df)		-1.94 NS (13df)		-4.33***(13df)	

Super-Scan values are means of six replicates
Other   "   "   "   "   three   "
NS = not significant; ** = p<0.01; *** = p<0.001

Nuclear magnetic resonance

Considerable interest has been shown in the last year or two in the application of modern instrumental methods to the analysis of meat components. Nuclear magnetic resonance (NMR) has been applied to both the measurement of hydroxyproline for collagen estimation,[27] and for protein estimation in meat.[28] In the first of these applications, ^{13}C pulsed Fourier transform NMR has been used to measure directly the hydroxyproline content of meat in comparison with the standard colorimetric procedure. The proton-decoupled ^{13}C-FTNMR spectrum was recorded at 25.03 MHz for a period of 30 min under standard conditions to ensure quantitative results for viscous samples of acid-hydrolised meat. Signals from the carbon atom at position 4 of the hydroxyproline molecule and from a suitable reference compound (the β-carbon atom of phenylpropanolamine HCl) occur in a clear part of the spectrum. The hydroxyproline content of a sample is calculated from the ratios of the peak heights multiplied by a constant derived from calibration graphs. Meat samples must be carefully defatted and then hydrolysed for 24 hrs, so there is no major saving of analysis time. In a collaborative exercise, good agreement was obtained between results for hydroxyproline content in a variety of sample types as determined by the standard colorimetric method in our laboratory and by FTNMR analysis at the Laboratory of the Government Chemist, London (Tables 2 and 3). The standard deviation of the FTNMR method was 0.16% at an hydroxyproline level of 3% w/w dried, defatted sample, compared with 0.10% for the colorimetric method at the 1.4% level.[27] The small differences in the results were within the acceptable limits for estimation of meat quality. Advantages of the NMR method are that the component of interest is measured directly, the response is linear at high hydroxyproline levels, and the possible inaccuracies of the colorimetric method are avoided.

Attempts were made to reduce the analysis time by altering the conditions of hydrolysis. It was found that a satisfactory FTNMR spectrum for semi-quantitative use could be obtained after digestion of a sample for 1.5 hr in 6M aqueous guanidinium chloride[30] in conjunction with 12N HCl at 100°C. Spectrometer analysis time was also reduced to 15 min, but interference was experienced at the C_4 resonance of hydroxyproline from incompletely degraded carbohydrate materials, a problem which did not occur with the full hydrolysis procedure. Whilst a ^{13}C-FTNMR-based procedure has certain merits, few laboratories will find it sufficiently attractive to justify allocation of spectrometer time for this purpose.

Less expensive NMR instruments have been produced, however, with routine analytical applications in mind, and the recent application of low resolution NMR to meat analysis is an example. The P-100 Protein Analyser uses low resolution (2.7 MHz) pulsed NMR to estimate the protein content of fresh meats.[28] The method is rapid and relatively simple; sample preparation is less demanding in that the protein does not need to be completely dissolved by lengthy digestion but only presented in finely dispersed form. The protein sample is mixed with an aqueous relaxation reagent containing a paramagnetic metallic ion. In this case the reagent is alkaline copper sulphate containing copper in the form of the cuprite ion $[Cu(OH)_4^{2-}]$ which can associate readily with the hydrogen nuclei of the water. Measurement of relaxation rates is achieved by pulsed NMR

Table 2 The 4-hydroxy-L-proline content of sections of beef muscles (as % w/w of fresh sample).

	by ^{13}C-FTNMR	by Colorimetry*
Longissimus Dorsi A	0.06	0.05
" " B	0.05	0.06
" " C	0.10	0.07
" " D	0.07	0.05
" " E	0.07	0.06
Complexus A	0.17	0.17
" B	0.16	0.17
Rhomboideus Thoracic A	0.16	0.11
" " B	0.18	0.12
" " C	0.17	0.12
Splenius Cervicis A	0.15	0.12
" " B	0.21	0.13

* Method of Stegemann and Stalder[29]

Table 3 The 4-hydroxy-L-proline content of meats and meat products (as % w/w of dried, defatted sample)

Sample	by ^{13}C-FTNMR	by Colorimetry*
Collagen	10.3	11.2
Brawn	3.2	3.7
Corned beef A	1.6	1.6
" " B	2.8	2.6
" " C	1.6	1.9
" " D	1.7	1.8
" " I	1.5	1.5
Canned ham A	1.7	1.5
" " B	1.4	1.5
" " C	1.6	1.4
Kidney (ox)	1.1	1.6
" (pig)	1.3	1.0
" (lamb)	0.5	0.6
Luncheon meat A	1.9	2.1
" " B	1.3	1.2
Salami	1.3	1.3

I.K. O'Neill et al[27] (selected data)
*Method of Stegemann and Stalder[29]

Table 4 Protein Estimation in meat by low resolution NMR

Sample	% protein by P-100			% protein by Kjeldahl		
	n	range	mean	n	range	mean
a) Calibration						
20-100 mg protein						
Dried beef (S)	7	81.0-85.7	82.5	2	80.4-81.0	80.7
Dried beef (Ld)	3	80.1-81.5	81.0	2	76.9-77.0	77.0
Fresh pork (Ld)	11	22.3-24.9	23.4	3	22.3-22.4	22.4
b) Calibration						
100-500 mg protein						
Fresh beef (Ld)	3	20.4-21.2	20.8	3	20.4	20.4
Fresh beef (S)	2	21.5	21.5	2	21.1-21.2	21.2
Frozen beef (Ld)	2	21.8-22.1	22.0	2	21.6-21.7	21.7
Fresh pork (Ld)	2	24.4-25.2	24.8	2	23.3-23.9	23.6

S = M. Sternomandibularis   Ld = M. Longissimus dorsi

techniques and any change in the state of the paramagentic ions will induce corresponding changes in the magnetic relaxation rate. If protein is added, the copper will, however, react preferentially with the peptide linkages (as in the biuret reaction) to form a complex which is only about one-eighth as effective as a relaxation agent as the original cuprite ion. Addition of protein, therefore, causes a linear reduction (within certain concentration ranges) in the relaxation rate such that the amount of protein present can be inferred directly from the difference in the relaxation rates of the sample and the reagent blank, provided that the system has been precalibrated by reference to conventional Kjeldahl analysis. Unfortunately, the alkaline copper reagent also shows a response to carbohydrate which, therefore, reduces its value for meat product analysis and limits it largely to fresh meat systems. Other relaxation reagents are available (acid copper, acid iron) which circumvent this problem but sensitivity to protein is reduced.

The results in Table 4 were obtained in our laboratories. In our hands, the P-100 tended to give values for dried, powdered beef which were higher and considerably more variable than those determined by Kjeldahl. Experiments with fresh pork also gave results which were similarly high and variable. Reproducibility and accuracy were improved to some extent for fresh meat samples by increasing the recommended sample size by a factor of five, with a concomitant increase in the final volume of copper reagent and homogenisation time required. No similar advantage was obtained with dried powdered samples because dispersion in the relaxation reagent was poor.

Throughout this paper, attention has been confined to a selection of techniques in which advances of relevance to meat research and the meat industry have occurred. It is not exhaustive and major topics have been omitted, but those covered may serve to remind or suggest new or alternative approaches to some problems encountered in this area of work.

REFERENCES

1.   O. Ouchterlony, Acta Pathol.Microbiol.Scand., 1948, 25, 186.
2.   C-B. Laurell, Anal.Biochem., 1966, 15, 45.
3.   R.R. Thompson, J.Assoc.Off.Anal.Chem., 1968, 51, 746.

4. K-P. Kaiser, G. Matheis, C. Kmita-Durrmann and H.-D. Belitz, Z.Lebensm.Unters.Forsch., 1980, 170, 334.

5. A.J. Sinclair and W.J. Slattery, Aust.Vet.J., 1982, 58, 79.

6. E. Engvall and P. Perlmann, Immunochemistry, 1971, 8, 871.

7. B.K. van Weeman and A.H.M.W. Schuurs, FEBS Lett., 1971, 15, 232.

8. E.K. Kang'ethe, S.J. Jones and R.L.S. Patterson, Meat Sci., 1982, 7, 229.

9. C.H.S. Hitchcock, F.J. Bailey, A.A. Crimes, D.A.G. Dean and P.J. Davis, J.Sci.Food Agric., 1981, 32, 157.

10. T. Kamiyama, Y. Katsube and K. Imaizumi, Jap.J.Vet.Sci., 1978, 40, 663.

11. D.J. Puckey, J.R. Norris and C.S. Gutteridge, J.Gen. Microbiol., 1980, 118, 535.

12. C.S. Gutteridge and D.J. Puckey, J.Gen.Microbiol., 1982, 128, 721.

13. H.L.C. Meuzelaar and P.G. Kistemaker, Anal.Chem., 1973, 45, 587.

14. H.L.C. Meuzelaar, P.G. Kistemaker, W. Eshuis and H.W.B. Engel, "Proc.2nd International Symposium on Rapid Methods and Automation in Microbiology", Learned Information Ltd, Oxford, 1977, p.225.

15. British Standard No.4401, Pt.II, 1969.

16. W.J. Olsman and C. Hitchcock, "Developments in Food Analysis Techniques", Applied Science Publishers Ltd, London, 1980, Vol.2, p.225.

17. D. Jones, D. Shorley and C. Hitchcock, J.Sci.Food Agric., 1982, 33, 677.

18. N.H. Poulter, W.R.D. Rangeley and R.A. Lawrie, Ann.Nutr. Aliment., 1977, 31, 245.

19. I. Hibbert and R.A. Lawrie, J.Food Technol., 1972, 7, 333.

20. N.H. Poulter and R.A. Lawrie, Meat Sci., 1980, 4, 21.

21. T. Hymowitz, J.W. Dudley, F.I. Collins and C.M. Brown, Crop Sci., 1974, 14, 713.

22. P.C. Williams, Cereal Chem., 1975, 52, 561.

23. W.G. Kruggel, R.A. Field, M.L. Riley, H.D. Radloff and M.H. Kristuna, J.Assoc.Off.Anal.Chem., 1981, 64, 692.

24. H. Martens, K.I. Hildrum, A. Bakker, S.A. Jensen, P. Lea, B. Eskeland, E. Vold and H. Russwurm, "Proc.26th European Meeting of Meat Research Workers", 1980, 1, 146.

25.   O-C. Bjarno, J.Assoc.Anal.Chem., 1981, 64, 1392.

26.   A. Bielby and D. Bartley, Meat Ind., 1983, 56, 43.

27.   I.K. O'Neill, M.L. Trimble and J.C. Casey, Meat Sci., 1979, 3, 223.

28.   L.R.H. Tipping, Meat Sci., 1982, 7, 279.

29.   H. Stegemann and K. Stalder, Clin.Chim.Acta, 1967, 18, 267.

30.   M.L. Jozefowicz, I.K. O'Neill and H.J. Prosser, Anal.Chem., 1977, 49, 1140.

11
The Chemistry of Vacuum and Gas Packaging of Meat

by Ted Hood

MEAT RESEARCH DEPARTMENT, THE AGRICULTURAL INSTITUTE, DUNSINEA
RESEARCH CENTRE, CASTLEKNOCK, CO. DUBLIN, EIRE

INTRODUCTION

Meat quality depends on a great many interrelated factors ranging from physical conditions of the processing and storage environment, biochemical properties of the muscle tissue, and microbiological status of the meat. The most serious problems associated with packaged fresh meat, which in turn significantly affect meat quality, are concerned with colour and colour stability and with water-holding capacity of the meat proteins. Chemical aspects of these parameters are therefore most relevant in the particular context of this paper. Earlier papers have already dealt with aspects of meat quality, such as tenderness and flavour, which are normally of much greater significance to eating quality of the meat. Differences in these latter quality attributes may be due to intrinsic properties of the meat which are not greatly affected by the method of packaging, or to extrinsic factors such as keeping quality which are microbiological in origin or due to some other cause. At the point of purchase, however, colour and colour stability and the absence of drip are of paramount importance.

This paper is concerned with the chemical aspects of these quality attributes of fresh meat in relation to various packaging conditions currently used commercially for meat storage and distribution.

MEAT PACKAGING

Meat marketing systems involving some form of packaging are nowadays quite diverse. At retail level they range from the ubiquitous supermarket prepacks in which meat is overwrapped in an air-breathing film, to the increasing use of high oxygen packs where the meat is placed in a rigid impervious container with sufficient gas volume to prevent significant reduction in oxygen content through metabolic activity during storage. Both these methods are designed to meet the technological demands of meat during retail storage and display.

The preferred method for the wholesale market is vacuum packaging, in which large primal cuts of meat are sealed under vacuum in impervious plastic bags. Following closure the bags are heat-shrunk to fit closely the contours of the meat contained therein. This is an excellent method of packaging fresh meat from a keeping quality point of view, since normal meat spoilage organisms are inhibited and the meat may be stored under refrigeration for several weeks without deterioration. The purple colour of the meat in anoxia precludes the widespread adoption of this method for retail marketing since the meat-buying consumer demands a bright-red colour at point of purchase.[1]

For wholesale cuts or primals, however, vacuum packaging is the universally preferred method. The essential features are that air is substantially removed during vacuum packaging and the meat is contained in an oxygen-impermeable film. Any remaining oxygen is converted to carbon dioxide by respiration of meat tissue and bacterial activity. A low concentration of oxygen in the residual atmosphere limits penetration into the surface of the meat to 1-2 mm. Approximately 1% oxygen may remain after prolonged storage.[2] The resulting thin layer of metmyoglobin which occurs on the surface does not, however, obscure the underlying myoglobin pigment.

Films used in vacuum packaging have a high degree of impermeability to oxygen and water vapour. Examples are coextruded EVA/PVDC/irradiated EVA shrink bags, polyvinylidene chloride (Saran), nylon/low density polyethylene and nylon/surlyn laminates. The Cryovac 'Barrier-Bag' sold by Grace, which is among the market leaders in fresh meat packaging, has an oxygen permeability of 20-30 $ml/m^2/24hr/atm$. and water vapour permeability of 0.4-0.5 $g/m^2/24hr/atm$. High vacuum (ca. 1 torr) gives the best results with respect to surface colour. High vacuum also improves both the appearance of fat, which otherwise may be discoloured by meat fluids, and the colour of the meat on final re-exposure to air.[3]

Flushing meat packages with inert gas avoids internal pressure on the meat due to vacuum packaging and provides an alternative system of anoxic storage which does not distort the shape of the packaged meat. This can be especially important in packaged meat in the form of small joints and steaks. Packaging retail cuts in oxygen-free nitrogen and/or carbon dioxide allows for extended storage because oxidative colour changes are prevented and bacterial growth is reduced.[4,5] The principal effect of carbon dioxide is of course microbiological rather than chemical, whilst nitrogen does not have any specific effect and may be considered to be quite inert.

Low levels of oxygen that may gain access during sealing or storage can cause oxidation, resulting in pronounced surface discolouration. Methods of removing low levels of oxygen using hydrogen in the gas together with a catalyst incorporated in the packaging film have been developed.[6,7] Retail cuts of fresh beef do not discolour during periods of up to 3 weeks at 0°C, using a catalytic oxygen scavenging system, and there is no difference between carbon dioxide and nitrogen, in relation to the rate of discolouration after eventual exposure to air.[8]

Returning to retail consumer packs, in this case the overriding consideration is the maintenance of a stable red colour and this is the principal criterion in designing a suitable package. For overwrapping, films must have good oxygen permeability whilst at the same time provide some barrier to the passage of water vapour. Landrock and Wallace[9] established that a packaging film must have a permeability of at least 5 $1/m^2/24hr/atm.$ to prevent oxidative browning. Commercially, however, an oxygen permeability value of 8-12 $1/m^2/24hr/atm.$ is more common.[10] The most widely used films for overwrapping include polyvinyl chloride (PVC), irradiated low-density polyethylene and ethylene/vinyl acetate copolymer. These combine the properties of high oxygen and low water vapour permeability with good stretch and shrink characteristics which provide a close fitting protective package during storage and retail display.

The advantage of an abundant supply of oxygen in developing a thick layer of red oxymyoglobin pigment at the surface is nowadays exploited commercially through the use of high oxygen packs. The process is the subject of several patents.[11,12] Earlier attempts to oxygenate slices of beef under pressure at 80 p.s.i. for 12 hr[7] before overwrapping in an impermeable film were only partially successful since oxidation proceeded as usual under these conditions. Holding similar slices in oxygen under high pressure maintained them in a fully oxygenated state for 12 days, but off-odours had developed after this time. Taylor and MacDougall[13] found that meat packaged in 80% oxygen and 20% CO_2 remained an attractive red colour for at least a week at +1°C. They concluded that the thick bright-red surface layer of oxymyoglobin masked the development of metmyoglobin in the underlying tissue. They pointed out that the environmental volume must be large enough to accommodate reduction by tissue respiration and prevent the oxygen pressure falling to an ineffective level during storage. During subsequent commercial trials these workers reported that colour deterioration during transport and storage was successfully delayed by holding in oxygen and the process was most effective in muscles prone to rapid discolouration.[14]

Several research workers have proposed the use of carbon monoxide in modified atmosphere systems for fresh meat storage.[15-17] Carbon monoxide combines with reduced myoglobin to form carboxymyoglobin. This bright-red derivative is more stable to oxidation than oxymyoglobin. El Badawi et al.[15] found that it was effective to flush packaged beef with 2% carbon monoxide before sealing in order to preserve the colour of fresh beef for 15 days. The amount of carbon monoxide bound to the meat pigment under these conditions is most unlikely to provoke any toxicological response within the blood of the consumer.[18]

PRE-SLAUGHTER STRESS

During post-mortem glycolysis the pH of normal tissue falls from an in vivo value of 7.2 to an ultimate value of 5.4-5.6. Both the rate of fall of pH and the ultimate pH value influence the colour, water-holding capacity and texture of the meat. Stress susceptible pigs are known to produce a high incidence of pale, soft exudative (PSE) meat. The condition is caused by an abnormally rapid fall in pH immediately after slaughter when the combined effects of low pH and high temperature in the muscle result in denaturation of sarcoplasmic and myofibrillar proteins[19,20] and there is also partial disruption of the sarcolemma.[21] PSE meat causes problems in packaging; because of its low water-holding capacity resulting from protein denaturation it produces excessive drip. The colour is also abnormally pale; due to denatured protein deposition the normal pink colour of muscle pigment is obscured;[22] the physical structure increases light scattering making the meat more opaque;[23] autoxidation is also increased causing colour fading.[24] Muscle glycolysis is relatively more rapid in pigs than bovines and PSE is rarely observed in beef animals.[25] Nevertheless there is evidence of appreciable differences in beef with respect to both colour and water-holding capacity due to differences in temperature/pH profiles post mortem.[21]

Biochemical conditions directly opposite to those producing PSE meat give rise to another type of abnormal meat, which is even more troublesome from a packaging point of view. This meat has a high pH, which is due to a low residual glycogen level remaining in the muscle at slaughter, following prolonged pre-slaughter stress. The problem is well known as dark-cutting or dry, firm and dark (DFD) beef. This meat is translucent and sticky to touch. It is unacceptable for retail packaging because of its dark-purple colour. Moreover, the high pH encourages the growth of putrefactive microorganisms, thereby significantly reducing the keeping quality of this meat.

These two extreme conditions produce unacceptable meat for packaging; PSE because of its extremely pale colour and excessive drip and DFD because of its dark colour, sticky texture and poor keeping quality. Between the extremes which produce these abnormal meat conditions, however, lie a wide range of meat types which are in general acceptable with respect to both colour and water-holding capacity but which nevertheless show considerable variability with respect to these two parameters.

MEAT COLOUR

Myoglobin is the principal pigment of fresh meat and the form that it takes is of prime importance in determining the colour of the meat. Haemoglobin is also present in meat though at a much lower level. Both proteins have similar extinction coefficients and at the same concentration contribute equally to colour at the surface.[24]

The myoglobin molecule consists of a haem nucleus attached to a globulin type protein component. Its molecular weight is about 17,000. The reactive haem group consists of a ring of four pyrollic nuclei coordinated with a central iron atom. Bonding of the iron to the tetrapyrollic ring structure satisfies four of the six coordination positions of the atom. The fifth is coordinated to an imidazole residue within the protein structure. The remaining sixth position is then available to bind a limited number of high-field ligands including oxygen. The oxygen binding properties of haem require that the fifth ligand must first be coordinated at one axial site on the iron; the basicity of the fifth ligand largely determines the equilibrium and kinetic characteristics of the sixth ligand binding to iron. The globin-histidine imidazole residue is especially effective as the fifth ligand, being a good π donor,[26] and as such it enhances the ability of ferrous-haem iron to form an electron-sharing addition compound with oxygen.[27]

The colour of fresh meat depends chiefly on the relative amounts of the three pigment derivatives of myoglobin present at the surface; reduced myoglobin (Mb), oxymyoglobin (MbO_2) and metmyoglobin (Mb^+). Reduced myoglobin, or deoxymyoglobin, is responsible for the purplish colour of freshly cut meat and meat held under anaerobic conditions, e.g. in a vacuum package. This myoglobin derivative is characterised by having an absorption peak at 555 nm in the green region of the spectrum. On exposure to the air myoglobin combines rapidly with oxygen forming bright-red oxymyoglobin which gives meat its typical attractive bright-red colour. Oxymyoglobin has relatively sharp maxima at 535-545 nm and 575-588 nm. When fresh meat is fully oxygenated the pigment is present 100% as oxymoglobin. Metmyoglobin is formed by oxidation to the ferric derivative. This pigment is brown with absorption bands at 505 and 630 nm and it is primarily responsible for the discolouration of meat, a condition which is characterised by a gradual darkening and browning of the meat surface. The accumulation of metmyoglobin is accompanied by a corresponding diminution in the proportion of oxymyoglobin. Ultimately the pigment may be totally converted to the oxidised form but this does not normally occur and in practice can only be achieved using an oxidative agent.

OXYGENATION AND OXIDATION

The colour cycle in fresh meat is reversible for all three major pigment derivatives and especially for the interconversion of deoxymyoglobin and oxymyoglobin. With its open sixth bonding site on haem iron, the reduced form of the myoglobin pigment, deoxymyoglobin, can bind high-field ligands such as oxygen, nitric oxide or carbon monoxide by π back bonding. Oxygen is obviously the most important of these in relation to fresh meat packaging. Nitric oxide forms nitrosylmyoglobin, the attractive red colour of uncooked cured meats. Carbon monoxide forms yet another red pigment derivative, carboxymyoglobin. It may have some application in modified atmospheres for meat storage.

Oxymyoglobin, the diamagnetic ferrous form of the derivative, is stable under high oxygen conditions. The ferrous derivative must be present for the oxygen to be found in a stable manner. The greater stability of oxygenated myoglobin is associated with maintaining the conformation of the protein.[28]

Due to the affinity of myoglobin for oxygen, oxymyoglobin forms rapidly on exposure to air. As oxygen partial pressure (pO_2) is lowered, however, and oxygen tension becomes low enough for partial deoxygenation to deoxymyoglobin, the stability of the system to oxidation is greatly diminished.[24]

George and Stratmann[29] found that the maximum rate of oxidation of myoglobin occurred at pO_2 1 to 20 mm of Hg depending on pH and temperature. Ledward[30] reported that the formation of metmyoglobin in fresh beef was maximal at pO_2 6 ± 3 mm and 7.5 ± 3 mm at 0°C and 7°C respectively. Metmyoglobin is the oxidised form of the myoglobin derivative in which iron is in the ferric (Fe^{3+}) state. The iron atom is covalently bound in metmyoglobin to a water molecule which in turn is bonded to a distal histidine.[28]

Autoxidation describes the spontaneous oxidation of oxymyoglobin which takes place slowly in the presence of oxygen. This reaction, however, cannot be envisaged as a simple transfer of one electron from the iron centre to a bound oxygen molecule since this reaction is thermodynamically unfavourable.[24] In fact this explains why myoglobin is an effective oxygen carrier, since oxygen binding would be only a fleeting process if the simple dissociation of oxygen from myoglobin were favourable to a one electron transfer. Given that oxygen must be reduced by two electrons and that only a single electron can be provided by ferrous myoglobin, the process is slow in chemical terms.[24] The half-life for autoxidation of bovine myoglobin at pH 6.5 at 22°C is 26.5 hr.[31] It should be noted that for packaged fresh meat storage times of interest to the meat retailer are often much longer than this.

A decreasing gradient (pO_2) occurs in meat and the maximal pO_2 for metmyoglobin formation occurs some distance below the surface for meat held in air. The actual depth of oxymyoglobin is determined by a number of factors including duration of exposure, temperature, oxygen tension, diffusion of oxygen through the tissue and its utilisation by the tissue.[32] Age of meat after slaughter also affects depth of penetration by oxygen and limits the thickness of the oxymyoglobin layer. Depth of penetration of oxygen for M. psoas major at 0°C after 48 hr exposure was 4.3 mm and 5.0 mm in samples taken at 3 and 20 days post mortem respectively. Corresponding values for the less respiratorily active M. longissimus dorsi were 4.9 mm and 7.11 mm.[33] Metmyoglobin which forms first at the interface between myoglobin and oxymyoglobin, where oxygen tension is optimal for oxidation, gradually diffuses outwards towards the surface.

Autoxidation is highly temperature dependent; Brown and Mebine[31] concluded a Q_{10} value $\approx$ 5. The reaction is also accelerated by low pH, which has been shown to reduce the stability constant for the haem-globin linkages.[34] An alternative possibility, with respect to pH, is that bound oxygen in oxymyglobin becomes protonated, oxygen can no longer be considered high-field, and one electron transfers from the iron, therefore becoming more energetically favourable.[24]

Metal ions also stimulate autoxidation of oxymyoglobin. In this respect Snyder and Skrdlant[35] found copper to be most active; iron, zinc and aluminium were found to be much less so.

The conditions necessary to prevent autoxidation, relevant to all meat packaging systems, may now be summarised. Low pO_2 must be avoided either by complete removal of oxymyoglobin by placing meat in anoxia or by exposure to high oxygen atmospheres. Temperatures should also be kept low and contamination by metal ions avoided. The high pH of dark cutting beef due to pre-slaughter stress would also be advantageous, although as we have already seen there are other problems associated with this condition.

Once oxidation has occurred the brown metmyoglobin permanently replaces the red oxymyoglobin, unless reducing conditions are present. These will be discussed in the next section. Denaturation of the protein (e.g. by increasing temperature, or lowering pH) results in haem becoming exposed. Binding by larger ligands (i.e. other than oxygen, nitric oxide or carbon monoxide) can then occur.[24] The denatured protein derivatives of myoglobin are ferrohaemochromes signifying that they are ferrous haem derivatives with a non-oxygen sixth position ligand. Several types of amino acid side chains can coordinate to the haem iron following denaturation of myoglobin. They may be derived from other proteins adjacent to myoglobin.[36] Ferrohaemochromes readily oxidise causing permanent brown discolouration of the meat.

Incident light (e.g. in a supermarket display case) is a contributory factor in the discolouration of fresh red meat.[26,37] This is particularly so of frozen meat, and the extent of the effect depends upon such factors as wavelength and intensity of light, temperature, oxygen pressure, meat pH, storage time, and in the case of frozen meat probably also electrolyte concentration and presence of free transition metal ions.[26] The effect of light is small under usual refrigerated conditions, provided UV wavelengths are avoided.[37] It is possible that sufficient singlet oxygen generated at oxygenated meat surfaces interacts in such a way that a modest increase in autoxidation is caused compared with comparable systems in the dark. The apparent complexity and low efficiency of the process probably explains why photocatalysed autoxidation is not a serious problem with fresh red meat.[26]

METMYOGLOBIN REDUCTION

Reducing systems present in meat were reported by Dean and Ball[38] to convert metmyoglobin during aerobic storage. Possible enzymic pathways for aerobic metmyoglobin reductase have been studied by many workers.[39-41] Metmyoglobin reduction in post rigor meat is primarily enzymic in nature. It may directly involve mitochondrial and/or submitochondrial particles as generators of reducing equivalents and also as scavengers of residual oxygen on vacuum packaging.[42] Hagler et al.[43] have recently reported the occurrence of a specific metmyoglobin reductase in beef heart muscle which rapidly and directly reduces metmyoglobin in vitro.

Loss of reducing activity in meat post mortem is due to a combination of factors including fall in tissue pH, depletion of

required substrates and co-factors, and ultimately complete loss
of structural integrity and functional properties of the
mitochondria.[42] The observation[44] that thorough mincing destroys
the reducing system in meat supports this view.

FACTORS IN FRESH MEAT DISCOLOURATION

Conditions which favour autoxidation, already discussed in
detail, favour discolouration of meat in all practical
situations. Temperature of storage exerts an appreciable
effect. High temperatures favour greater oxygen scavenging by
residual respiratory enzymes as well as other oxygen-consuming
processes such as fat oxidation, thereby leading to the low pO_2
levels which are conducive to autoxidation.[18] Low temperature,
on the other hand, promotes increased penetration of oxygen into
the surface and oxygen solubility in tissue fluids is enhanced,
both increasing the depth of oxymyoglobin at the surface. The
dissociation of oxygen from oxymyoglobin is also enhanced by
higher temperatures thereby increasing the tendency for
autoxidation of the deoxygenated myoglobin produced.[18]

Intermuscular variability is another major factor of
appreciable practical importance.[37] Table 1 shows the results of
an analysis of variance of discolouration data for four muscles
from ten animals at three temperatures 0, 5, 10°C.

Table 1 Components of variance for discolouration data

Effect	Total variation percent	F-test significance
Animal	7.3	p <0.001
Muscle	45.5	p <0.001
Temperature	32.5	p <0.001
Muscle X temperature	3.5	p <0.05
Error 1	12.0	
Total	100.0	

The results are typical for a wide range of beef. Effects
due to animal, temperature, muscle and muscle/temperature

interaction are all significant. The muscle effect is most important, accounting for almost half of the total variation in this case. Stability to oxidation is characteristic of the individual muscle and the profile of intermuscular difference is consistent within the carcass. Thus M. psoas major invariably shows poor colour stability; conversely M. longissimus dorsi and M. semimembranosus are stable; whilst M. gluteus medius has intermediate colour stability. The fact that the least stable colour occurs in some of the most expensive cuts creates obvious problems in relation to meat packaging operations. Discolouration measurements in different muscles from the same animal are positively interrelated, the highest correlation being between M. gluteus medius and M. semimembranosus.[37] Discolouration data for four muscles M. longissimus dorsi, M. psoas major, M. semimembranosus and M. gluteus medius after 96 hr at 0, 5 and 10°C are shown in Fig.1.[37] The intermuscular effect is clearly evident. The degree of discolouration is almost eight times greater in M. psoas major than in M. longissimus dorsi at 0°C after 96 hr storage. The effect of temperature is also important in relation to rate of discolouration. The degree of discolouration in 96 hr at 10°C is two to five times that at 0°C depending on the muscle.

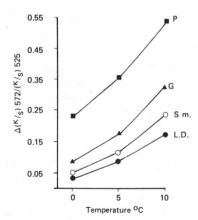

Figure 1 Effect of temperature on rate of discolouration in four bovine muscles[37] (n = 10).

The principal biochemical differences between the least stable and most stable muscles, M. psoas major and M. longissimus dorsi respectively, have been compared.[33] A summary of the major differences is given in Table 2.

Table 2 Biochemical properties in M. psoas major and M. longissimus dorsi.

	M.psoas major	M.longissimus dorsi
Oxygen consumption rate (OCR)	High	Low
Oxygenation of intact muscle	Efficient	Less efficient
Oxymyoglobin layer	Narrow	Wide
Succinic dehydrogenase (SDH) activity	High	Low
Oxidation with $K_3Fe(CN)_6$	Easy	Difficult
Myoglobin content	Lower	Higher
Conversion of oxymyoglobin to myoglobin	Rapid	Slow
Metmyoglobin formation during conversion of oxymyoglobin to myoglobin	Strong	Weak
Metmyoglobin reducing activity (MRA)	Low	High
Colour stability	Poor	Good

M. psoas major is characterised by having a rapid rate of oxygen consumption and efficient oxygenation or 'blooming' on exposure to air. It also loses oxygen readily when the oxygenated muscle is placed in anoxia and reduction to deoxymyoglobin is rapid. This muscle is also susceptible to oxidation and metmyoglobin tends to form during the reduction reaction. Its high respiratory rate is associated with high SDH activity but with a relatively low myoglobin content. Conversely, M. longissimus dorsi, which has good colour stability, has a slower oxygen consumption rate and takes longer to bloom on exposure to air. Conversion to myoglobin is relatively slow. It resists oxidation more effectively and does not tend to form metmyoglobin during conversion to deoxymyoglobin.

Aged meat develops a better colour when cut, compared with non-aged meat. With the exception of very fresh meat[37,45] aged meat discolours more rapidly the longer the period of storage. [8,46] Both the better blooming and faster rate of discolouration of aged meat result from diminution of the meat's enzymic activity which occurs during the conditioning period.[46] A thicker layer of oxymyoglobin forms in aged meat because the rate of oxygen consumption is lowered as substrates of glycolysis are exhausted, allowing greater penetration of oxygen into the meat. Metmyoglobin formed in the region of low pO_2 is no longer converted to deoxymyoglobin because reducing intermediates, particularly NADH, are no longer being formed.[33]

WATER-HOLDING CAPACITY

Skeletal muscle contains about 75% water and the strength with which it is bound by muscle proteins is of great importance to the quality of meat and meat proteins. In fresh meat packaging the appearance of free water in the form of a blood-like fluid, or 'drip', is very unsightly even though the actual amount is often quite small (<2%). Packages are designed to minimise the effect but the occurrence of drip depends on intrinsic properties of the meat so that package design alone cannot overcome this problem.

According to Hamm[47] the myofibrillar proteins actin and myosin are primarily responsible for the binding of water in muscle. About 4-5% is tightly bound as hydration water which is hardly influenced by changes in the structure and charges of muscle proteins. The remaining 95% of water in muscle, however, is concerned with changes which occur in water-holding capacity. In the model proposed by Hamm, the water is considered to be immobilised, partly by charged groups on adjacent myofibrillar protein molecules and partly by the physical configuration of these proteins and the physical barriers created by the sarcolemma and other membrane sheaths, such as the sarcoplasmic reticulum. The amount of water immobilised within the tissue is influenced by the spatial molecular arrangement of the myofibrillar proteins.[47] Any factor which causes an enlargement of the myofibrillar network such as adjustment of pH away from the isoelectric point, causing increased electrostatic repulsion between charged groups, will allow more water to be immobilised.

A tightening of the myofilament structure on the other hand, such as adjustment of pH towards the isoelectric point, will result in water being less efficiently immobilised and more easily expressed from the meat under a variety of physical forces. The water-holding capacity of meat is at a minimum at pH 5, and this corresponds to the isoelectric point of actomyosin which makes up the bulk of the structural muscle proteins. This is also close to the normal ultimate pH of fresh meat.[47] The mechanism of water-holding in meat is discussed in detail by Offer in Chapter 5.

The rate of fall of pH <u>post mortem</u> has already been discussed in relation to the occurrence of PSE meat and the influence of pH/temperature profiles as contributory factors to variability in colour and drip loss have been pointed out. Although the PSE condition is rarely observed in beef carcasses[25] substantial variation in the rate of pH fall occurs between and within muscles which results in significant differences in both these meat quality attributes.[21] Variations in pH fall in beef carcasses are due partly to inherent differences in muscle metabolism and function[48] and partly to differences in cooling

Table 3 Denaturation of sarcoplasmic and myofibrillar proteins and amount of drip loss, as a function of the depth in the chilled beef round. Measurements were made in <u>M. adductor</u> and <u>M. semimembranosus</u> two days after slaughter.[49]

Depth in carcass (cm)	Myofibrillar ATPase activity (μmol Pi/mg protein/ min x 10^2)	CPK[a] activity (units/g x 10^{-2})	Percent drip (300g, 30 min)
<u>AD</u>			
1.5	7.1	53	10.4
5	6.3	37	18.4
8	4.4	10	21.7
<u>SM</u>			
1.5		55	6.9
5		40	14.3
8		15	19.4

a Creatine phosphokinase

rates within the muscles of the carcass.[21] The faster rate of pH fall in the deep, slow cooling parts of conventially-chilled beef carcasses results in increased denaturation of myofibrillar and sarcoplasmic proteins and higher drip loss[21] (Table 3).

The effect due to differing pH/temperature profiles may readily be observed in a cross-secion of the beef round. The deep meat adjacent to the femur is lighter in colour, softer in texture and wetter compared with meat closer to the surface.[21] Whilst this degree of variation, occurring sometimes within a single muscle, poses no particular problem in a traditional butchery operation, it is a significant contributory factor to the variability in colour and drip loss between individual packages of meat. Major differences not only detract from overall appearance but may determine ultimate acceptability to the consumer. The development of central prepackaging is also inhibited because of these difficulties, particularly widespread distribution such as an export orientated operation.

HOT BONING AND ELECTRICAL STIMULATION

Fabrication of beef into primal cuts before chilling followed by vacuum packaging of the meat has received a lot of attention in recent years. Hot boning, as this process is called, involves removal of the pre-rigor meat, followed by controlled chilling of the vacuum packaged cuts. The process provides a means of avoiding the large temperature gradients which occur when carcasses are chilled in the traditional manner.[50,51] It is important under these conditions to avoid cold-induced muscle contraction of the boned out muscle, since this is associated with toughening of the meat. This can be achieved by holding the meat above 10°C until rigor mortis is completed.[52]

Electrical stimulation of carcasses soon after slaughter is now widely practised by industry to accelerate post-mortem glycolysis and the onset of rigor mortis. This enables rapid cooling to be begun much sooner after slaughter without the danger of cold shortening and component toughening of the meat when it is cooked.[53] Other advantages claimed for electrical stimulation include acceleration of the natural tenderisation of meat which occurs during ageing, reducing evaporative weight loss and accelerating the so-called blooming of meat.[53] On the other hand, electrical stimulation may result in increased protein denaturation which results from accelerated rate of glycolysis and high temperature. Despite this, Bendall[53] emphasised that he was unable to detect any significant increase in drip losses during storage after electrical stimulation of beef carcasses.

Tarrant and Mothersill,[49] however, clearly show increased drip loss associated with accelerated rates of glycolysis. Minimum protein denaturation and maximum water-holding capacity were reported when the temperature fell to 15°C or below before the pH reached 6.0. In hot-boned M. semimembranosus Tarrant[51] found these conditions obtained throughout the muscle, indicating a substantial improvement in water-holding capacity as a result of hot boning. More recently Taylor et al.[54] found that hot boning followed by delayed chilling reduced drip loss in vacuum packaged cuts; use of electrical stimulation resulted in a smaller reduction in drip loss. Another effect of hot boning was to produce a more uniform colour in large muscles which normally cool unevenly in the intact side. Electrical stimulation again reduced the effect.

REFERENCES

1. D.E. Hood and E.B. Riordan, J.Food Technol., 1973, 8, 333.

2. A.A. Taylor and B.G. Shaw, J.Food Technol., 1977, 12, 515.

3. S.C. Seideman, Z.L. Carpenter, G.C. Smith and K.E. Hoke, J.Food Sci., 1976, 41, 732.

4. D.L. Huffman, J.Food Sci., 1974, 39, 723

5. W. Partmann and H.K. Frank, Proc.13th Int.Cong.Refrig., Washington, D.C., 1971, 3, 17.

6. L.J. Ernst, "Proc.28th Ann.Meet.Res.Develop.Assoc.Milit. Food Pack.", Systems Inc., Chicago, 1974, p.78.

7. G.L. Zummerman and H.E. Snyder, J.Food Sci., 1969, 34, 258.

8. M. O'Keeffe and D.E. Hood, Meat Sci., 1980-81, 5, 27.

9. A.H. Landrock and G.A. Wallace, Food Technol., 1955, 9, 194.

10. S.S.H. Rizvi, CRC Crit.Rev.in Food Sci.and Nut., 1981 (Feb.), 111.

11. D.L. Georgala and C.M. Davidson, British Patent 1 199 998, 1970.

12. Schweisfurth and Kalle Aktiengesellschaft, British Patent 1 186 978, 1970.

13. A.A. Taylor and D.B. MacDougall, J.Food Technol., 1973, 8, 453.

14. D.B. MacDougall and A.A. Taylor, J.Food Technol, 1975, 10, 339.

15. A.A. El Badawi, R.F. Cain, C.E. Samuels and A.F. Anglemeier, Food Technol., 1964, 18, 159.

16. D.S. Clark, C.P. Lenz and L.A. Roth, Can.Inst.Food Sci. Technol.J., 1976, 9, 114.

17. R.E. Woodruff, U.S. Patent 3 930 040, 1976.

18. C.L. Walters, "MEAT", Butterworths, 1975, p.385.

19. R.K. Scopes, Biochem.J., 1964, 91, 201.

20. I.F. Penny, Biochem.J., 1967, 104, 609.

21. P.V. Tarrant, "Food Proteins", App.Sci.Publishers, London, 1982, p.261.

22. J.V. McLoughlin and G. Goldspink, Nature, 1963, 193, 584.

23. D.B. MacDougall, J.Sci.Food Agric., 1970, 21, 568.

24. D.J. Livingstone and W.D. Brown, Food Technol., 1981, 5, 244.

25. M.C. Hunt and H.B. Heddrick, J.Food Sci., 1977, 42, 716.

26. G.G. Giddings, J.Food Sci., 1977, 42, 288.

27. J.M. Rifkind, "Inorganic Biochemistry", Elsevier Sci. Publ.Co., 1973, Vol.2.

28. S. Govindarajan, CRC Crit.Rev. Food Technol., 1973, 4, 117.

29. P. George and C.J. Stratmann, Biochem.J., 1952, 51, 418.

30. D.A. Ledward, J.Food Sci., 1970, 35, 33.

31. W.D. Brown and L.B. Mebine, J.Biol.Chem., 1969, 244, 6696.

32. D.B. MacDougall, "Proc.Symp.Meat Chilling," MRI, Bristol, 1972, p.8.1.

33. M. O'Keeffe and D.E. Hood, Meat Sci., 1982, 7, 209.

34. C. Fronticelli and E. Bucci, Biochim.Biophys.Acta, 1963, 78, 530.

35. H.E. Snyder and H.B. Skrdlant, J.Food Sci., 1966, 31, 468.

36. D.A. Ledward, J.Food Sci., 1971, 36, 883.

37. D.E. Hood, Meat Sci., 1980, 4, 247.

38. R.W. Dean and C.O. Ball, Food Technol., 1960, 14, 271.

39. C.L. Walters and A. Taylor, Food Technol., 1963, 17, 354.

40. A.J. Cutaia and Z.J. Ordal, Food Technol., 1964, 18, 757.

41. B.M. Watts, J. Kendrick, M.W. Zipser, B. Hutchins and B. Saleh, J.Food Sci., 1966, 31, 855.

42. G.G. Giddings, CRC Crit.Rev. Food Techol., 1974, 5 (2), 143.

43. L. Hagler, R.I. Coppes Jr. and R.H. Herman, <u>J.Biol.Chem.</u>, 1979, <u>254</u> (14), 6505.

44. D.A. Ledward, C.G. Smith, H.M. Clarke and M. Nicholson, <u>Meat Sci.</u>, 1977, <u>1</u>, 149.

45. M. O'Keeffe and D.E. Hood, <u>Meat Sci.</u>, 1980-81, <u>5</u>, 267.

46. D.B. MacDougall and D.N. Rhodes, <u>Proc.Pack & Pres.Food Int. Congress</u>, Wembley, 1972.

47. R. Hamm, "MEAT", Butterworths, 1975, p.321.

48. J.R. Bendall, <u>Meat Sci.</u>, 1978, <u>2</u>, 91.

49. P.V. Tarrant and C. Mothersill, <u>J.Sci.Food Agric.</u>, 1977, <u>28</u>, 739.

50. M.J. Follett, C.A. Norman and P.W. Ratcliff, <u>J.Food Technol.</u>, 1974, <u>9</u>, 509.

51. P.V. Tarrant, <u>J.Sci.Food Agric.</u>, 1977, <u>28</u>, 927.

52. J.R. Bendall, "Proc.Symp.Meat Chilling," MRI, Bristol, 1972, p.3.1.

53. J.R. Bendall, "Developments in Meat Science - 1", Appl.Sci. Publ., London, 1980, p.37.

54. A.A. Taylor, B.G. Shaw and D.B. MacDougall, <u>Meat Sci.</u>, 1980-81, <u>5</u>, 109.

12
The Chemistry of Meat Binding

By Glenn R. Schmidt and Graham R. Trout
DEPARTMENT OF ANIMAL SCIENCES, COLORADO STATE UNIVERSITY, FORT COLLINS, COLORADO 80523, U.S.A.

INTRODUCTION

Meat components perform many functions in processed products. Intact muscle pieces, fibres and myofibrils contribute directly to texture. Texture is also affected by the inclusion of connective tissue and adipose tissue. The inclusion of air, water droplets and melted fat within the product affect the texture of the cooked material. The species, composition and anatomical location affect meat ingredient functionality.

Although different types of meat products, such as emulsified, particulate and sectioned and formed meats, vary greatly in their method of preparation and meat particle size, they do have one common characteristic. This is, their ability to bind their constituent meat components together so as to form a cohesive product. The strength of this binding is important because it determines, to a large degree, the quality of the products. The same basic mechanisms of binding apply to most meat systems.[1]

DEFINITION OF BINDING STRENGTH

As referred to here, binding strength is defined as the force per unit cross-sectional area required to pull apart bound pieces of meat. As such it includes a measure of both the cohesive force exerted between the binding matrix and the meat pieces and the strength of the binding matrix itself.

The mechanism of binding in meat systems is very complex in nature and is not fully understood. However, the main factors that determine the efficacy of binding are protein extraction, mechanical treatment, presence and concentration of added salts and temperature of heating.[1,2,3] These may not be the only factors determining binding ability, but they are the ones that are generally accepted as being the most important and will be covered in detail here.

PROTEIN EXTRACTION

Bard[4] reviewed some factors influencing the extractability of salt soluble proteins from muscle tissue and concluded; (a) temperatures in the range of -5°C to 2°C gave maximum protein extraction, (b) increasing extraction time increased protein extraction, up to an extraction time of 15 hr, (c) pre-rigor meat is more extractable than post-rigor meat, and (d) a sodium chloride concentration of 10% extracted the most protein. Gillett et al.[5] determined the total amount of protein extracted from muscle, under varying conditions, and concurred with Bard on most points except for the temperature of maximum extraction.

Table 1 Structural components of meat products.

Structures:

        muscle pieces, fibres and myofibrils

        connective tissue

        collagen fibrils

        adipose tissue, cells and lipid droplets

        air

        textured plant proteins

Coatings of Structures:

        soluble myofibrillar and sarcoplasmic proteins

        non-meat proteins - plant, milk, plasma

        true emulsions

Matrix Formation:

        actomyosin-myosin aggregates

        plasma

        bound water

Inclusions:

        fat droplets

        free water

        plant proteins

Effectors:

        ions

        mechanical

        temperature

        pH

        rigor

They found the optimum temperature for extraction was 7.2°C with a marked decrease in extractability at 0°C, the temperature which Bard found was optimum. However, this difference may be attributed to the fact that the conditions of extraction were not the same for both groups. Gillett used a shorter extraction time (6 min vs. 30 min), a higher salt concentration (6% vs. 3.9%) and a higher solvent to meat ratio (3:1 vs. 2:1).

The effect of vacuum and extraction time on the extractability of crude myosin from pre- and post-rigor meat was the object of a study by Solomon and Schmidt.[6] They concluded; (a) the extraction of crude myosin increased linearly with extraction time, (b) vacuum increased the amount of crude myosin extracted by 20% over that of the non-vacuum treatment, and (c) 65% more crude myosin was extracted from pre-rigor meat than from post-rigor meat. These results with pre-rigor meat are in good agreement with those of Saffle and Galbreath[7] and Acton and Saffle[8] who both found that the amount of salt soluble protein extracted from pre-rigor meat was 50% greater than that extracted from post-rigor meat.

Saffle and Galbreath[7] studied the effect of increasing the pH of muscle from pH 5.5 to 6.5 in small increments on protein extraction. Increasing pH increased amounts of salt soluble protein that were extracted. Both Acton and Saffle[8] and Awad et al.[9] found that freezing meat reduced the amount of total protein and salt soluble protein extracted. The effect of polyphosphates on the extractability of myofibrillar proteins and crude myosin was studied by Turner et al.[10] using sodium tripolyphosphate and by Hamm and Grabowska[11] using tetrasodium pyrophosphate. Both groups found that in the presence of salt, polyphosphate increased the amount of protein that could be extracted, over that of salt alone.

The role of solubilised meat proteins is to bind to the insoluble components in the protein matrix to form a coherent stable combination with each other.[12] This concept appears quite compatible with much of the research that has been done in the area of relating protein solubilisation to binding in meat products. Siegel and Schmidt[13] found that increasing the amount of extracted myosin between meat surfaces produced a linear increase in binding strength in a model binding system. Using a meat system similar to that used in weiners, Randall[14] found that when as little as 2.5% of the total protein was replaced with salt extracted proteins, the binding strength of the resultant product increased. When investigating the importance of particle size on the binding in poultry rolls, Acton[15] found that reducing the particle size increased the amount of salt soluble protein extracted. With the increase in protein extraction there was a concurrent increase in binding strength, and the relationship between them was significant and highly correlated (r=0.91). A study by Theno et al.[16] found that increasing the massaging time of hams increased the amount of protein extracted in a hyperbolic fashion, with the curve approaching an assymtote at 14% protein. Although this was the case, the binding strength only increased with protein content up to 12% protein, after which it remained fairly constant. This observation may be explained by the fact that increased massaging time caused fibre disruption and weakening which counteracted any increase in binding strength produced by the additional protein extracted. When investigating the effect of storage time on the functionality of frozen meat (pork and beef), Miller et al.[17] found that increased storage time reduced the total amount of protein that could be extracted from meat. They found significant correlations between reduction in total extractable protein and reduction in binding strength of weiners, for both beef and pork, with the correlation for pork being greater than that of beef (r=0.93 cf. r=0.83).

MECHANICAL TREATMENT

Mechanical treatment is one characteristic that is applied to most processed meat products and is a prerequisite for effective binding. There are many types of mechanical treatments that can be applied to meat to improve its binding characteristics. Methods commonly used to increase binding in sectioned and formed meat products are mixing, massaging, tumbling and mechanical tenderisation. The importance of mechanical action has led to the study of this topic to try to ascertain the underlying mechanism that enhances binding.

The work of Schnell and others[2] clearly showed, by measuring the amount of nucleic acid in the cookout, that mixing causes cell disruption and breakage with subsequent release of the cell contents including, presumably, the myofibrillar proteins. Siegel et al.[18] showed that increased massaging time increased the amount of protein extracted, hence this work agrees with that of Schnell and workers[2] in that increased protein extraction would be expected with increased cell disruption. Koo[19] showed a similar effect occurred with tumbling. He found that there was more myosin extracted from tumbled hams than from non-tumbled hams. Using light microscopy Theno et al.[16] were able to show conclusively that increased massaging time increased both fibre disruption and the amount of myofibrillar protein solubilised. The increased myofibrillar protein solubilisation may not be the only important contribution of cell disruption as Schnell et al.[2] noted that addition of RNA, which was also released by mixing, increased the binding strength of poultry rolls. Another important role of mechanical treatment was illustrated by Theno and workers,[20] using the scanning electron microscope. They showed that myofibrils and muscle fibres, which are both normally tightly packed, separate after massaging. Once this structure is opened the solubilised proteins of the exudate are worked into the loose fibre structure allowing a more cohesive bond to form between the protein matrix and the meat surface. This concept is very similar to that described by Kotter and Fischer[12] for emulsion products. They theorise that the solubilised protein binds to the insoluble fibres which have been separated and dispersed by chopping.

Both Maesso et al.[21] and Schnell et al.[2] showed that mixing for 3 min increased the binding strength of poultry rolls when compared to non-mixed rolls. Pepper and Schmidt,[22] using a more quantitative approach with beef rolls, found that progressively increasing the mixing time from 5 min to 30 min produced corresponding increases in binding strength. But when the mixing times exceeded 30 min the binding strength did not increase with time and in some cases it actually decreased. Increasing the length of time hams were massaged from 0 hr to 24 hr increased the binding strength of the resultant products only up to a massaging time of 12 hr.[18] Beyond a massaging time of 12 hr, the binding strength of the hams started to decline, which is a comparable result to that obtained by Pepper and Schmidt[22] with beef rolls. This reduction in binding strength produced by extended mechanical treatment may be explained by the fact that with long periods of mechanical treatment the muscle fibres become excessively disrupted and the meat starts to lose its structural integrity. Krause et al.[23] investigated the effect of different tumbling treatments on sectioned and formed canned hams. They concluded that total tumbling times of 95 min and 180 min improved the binding strength of canned hams, over those hams not tumbled, and that intermittent tumbling had no advantage over continuous tumbling as long as the total tumbling time was the same. The importance of a certain degree of meat fibre disruption for improved binding was pointed out by McGowan[24] in his patent. He found that fraying at the meat surface, without disruption of the integrity of the meat, allows a binding matrix to adhere more strongly to the meat surface and hence increase binding strength of the product.

THE PRESENCE AND CONCENTRATION OF ADDED SALTS

There is a range of salts that may be added to meat products, but because of taste and toxicological considerations the two most widely used are sodium chloride and the sodium salts of the polyphosphoric acids. The main role of salts is to contribute to the ionic strength of the system, with an additional role being to alter the pH. Alkaline polyphosphates tend to increase the pH, usually by 0.1 to 0.4 units depending on type and concentration, while sodium chloride and other neutral salts tend to reduce the pH by 0.1 to 0.2 pH units.[25] The effect of polyphosphates on pH is due to their alkaline nature, while that due to sodium chloride has been theorised to be due to the displacement of hydrogen ions by sodium ions on the meat protein surface, with the liberated hydrogen ions producing the drop in pH. The mechanisms by which salts increase the binding ability of a protein matrix are:

(a) by increasing the amount of protein extracted;

(b) by altering the ionic and pH environment such that the resultant heat set protein matrix forms a coherent three-dimensional structure.

Grabowska and Sikorski[26] studied the effect of changing pH and salt concentration on the gel strength of washed fish myofibrils and extracted myofibrillar proteins. They found that increasing the pH incrementally from 5 to 7.5 produced a corresponding increase in gel strength. When the salt concentration was increased from 0 to 5%, the gel strength increased up to a salt concentration of 3% (0.5 μ), after which increasing salt concentration had little beneficial effect. When investigating the effect of ionic strength and pH on the strength of heat set myosin gels, Ishioroshi et al.[27] found pH 6.0 and an ionic strength of 0.2 produced the maximum gel strength. These results seem to contradict those of Grabowska and Sikorski, but the difference in results may be explained in terms of the work of Yasui et al.[28] who showed that the presence of bound actin affected the gel characteristics of myosin. Mixtures of actin and myosin exhibited maximum gel strength at pH 6.0 and an ionic strength of 0.7. An explanation of how salts increase gel strength was put forward by Siegel et al.[29] who showed, using scanning electron microscopy, that when myosin and actomyosin were heated in high ionic strength salt solutions, the proteins formed a coherent three-dimensional network of fibres. In the absence of added salts the same proteins formed a spongy structure with little strength. From this they concluded that the characteristic three-dimensional structure produced by the addition of salts, was necessary for the meat proteins to produce a satisfactory binding strength. Yasui et al.,[28] also using scanning electron microscopy, found that the same type of structure occurred for both myosin and actomyosin at pH 6.0 and ionic strength of 0.6, but not with actin.

Turner et al.,[10] in investigating the effectiveness of isolated crude myosin preparations as binding agents between meat pieces, found that increasing the concentration of salt in the preparation (from 0 to 1.0 μ) produced an increase in binding strength. Using the same procedure as Turner et al.,[10] Siegel and Schmidt[13] looked at the effect of added polyphosphate (0.5%), change of pH (6.0, 7.0 and 8.0) as well as added salt (0%, 2%, 4%, 6%) on the binding strength of crude myosin. Their results for the effect of salt on binding strength were essentially the same as those obtained by Turner and workers. In addition, they found that the presence of 0.5% sodium tripolyphosphate produced a significant increase in binding strength, but unexpectedly the pH changes had no significant effect. The absence of any significant effect of pH on binding ability may be explained if it is considered that the meat present in the system buffered the myosin preparations so that they all had similar final pH's.

THE TEMPERATURE OF HEATING

As pointed out by Schnell et al.[2] and Vadehra and Baker,[3] the binding between meat pieces is a heat initiated reaction, since no binding occurs in the raw state. Kotter and Fischer[12] suggested that heating caused the previously dissolved proteins to rearrange so that they could interact with the insoluble proteins on the meat surface and in so doing form a cohesive structure. The work of Hamm and Deatherage[30] indicates that this process begins at 45°C and describes the molecular interaction involved as non-covalent. The latter conclusion was based on the fact that even heating to 70°C caused no observable formation of intermolecular disulphide bonds. Hamm[31] further postulated that the stabilising bonds formed after heat denaturation were mainly hydrogen and ionic interactions.

Yasui et al.[28] investigated the effect of temperature on rigidity of rabbit myosin and actomyosin gels. They found that the gel strength started to increase at 40°C and reached a maximum at 60°C. A similar result was obtained by Grabowska and Sikorski[26] using fish myofibrils, with the difference being that the increase in gel strength started at 30°C and continued up to a temperature of 80°C. These results are confirmed in principle by the work of Quinn et al.[32] who showed, using differential scanning calorimetry, that denaturation of meat (beef) proteins begins at about 50°C and continues with increasing temperature up to 90°C. This work and that of Wright et al.[33] shows that the temperature range of denaturation of the different protein components was a characteristic of the species of animal the protein came from, the pH and the ionic strength. This information would tend to explain the varying results obtained by different workers in this area, as each group used different combinations of animal species, pH and ionic strength.

Acton[15] investigated the effect of temperature of cooking on the binding ability of poultry meat. The results obtained on binding ability were basically the same as those obtained by workers measuring gel strength of isolated proteins. In essence, the binding strength started to increase at 40°C and reached a maximum at 80°C after which it decreased slightly with temperature up to 100°C. The effect of temperature on the binding ability of crude myosin (beef) was investigated by Siegel and Schmidt.[13,29] Their results showed that binding strength started to increase at 55°C and then increased linearly with temperature to 80°C, but did not show the same decrease in binding ability after 80°C as was previously reported by Acton.[15] The difference, in response of binding ability to temperature, obtained by the two different groups of workers may, like the variation in gelling of purified proteins, be due to difference in species of animal, pH and ionic strength.

From the research on the effect of temperature on gel strength and binding ability of meat proteins it can be concluded that there is a relationship between the temperature of heating and the presence and concentration of different salts. The exact relationship has not been clearly elucidated, but the implication is that the temperature at which maximum binding occurs is dependent on the presence of specific salts and hence the ionic strength and pH.[32]

ROLE OF SPECIFIC MEAT PROTEINS IN BINDING

Of the three main groups of proteins found in muscle, sarcoplasmic, myofibrillar and stromal, the myofibrillar proteins are the ones that have been implicated as being the most important in binding. However, there is some evidence which indicates that both the sarcoplasmic proteins and the stromal proteins contribute significantly to the binding characteristics of meat.

Sarcoplasmic proteins

The importance of sarcoplasmic proteins in the binding of meat pieces in a poultry loaf was investigated by Acton and McCaskill.[34] They removed 35% of the sarcoplasmic proteins, mainly from the meat surface, by washing the pieces with deionised water. After this they measured the binding ability of the washed meat pieces when used in a poultry loaf, either with or without 2% added salt. They found that the binding ability of the washed meat cubes, both in the presence and absence of 2% salt, did not differ significantly from that of the unwashed control meat pieces of similar protein content. Hence, in spite of the increased percentage of myofibrillar proteins at the meat surface, produced by the removal of sarcoplasmic proteins, there was no increase in binding ability. This is contrary to what would be expected if the sarcoplasmic proteins contributed little

or nothing to the binding ability of the meat, and illustrates
the positive effects sarcoplasmic proteins have on binding.
Siegel and Schmidt[13] obtained similar results when measuring the
binding ability of washed and unwashed beef muscle homogenates,
of equal protein content (5%), in a model binding system. They
found that, even with complete removal of the sarcoplasmic
proteins, there was no significant increase in binding ability in
the presence of 6% salt and 2% sodium tripolyphosphate. Both
MacFarlane et al.[35] and Ford et al.[36] investigated the ability of
extracted sarcoplasmic proteins, used separately or in
conjunction with extracted myosin, to bind meat pieces in both a
model binding system and a restructured beef steak. The general
conclusion drawn from the results of this group of workers is
that when the ionic strength of the binding matrix is low (below
0.4 μ), the sarcoplasmic proteins make a significant contribution
to the binding ability of the system. But, when the ionic
strength is increased beyond 0.4 μ, sarcoplasmic proteins have
little beneficial effect on binding ability and, in some cases,
the effect is detrimental. This was only true when the
sarcoplasmic proteins were used with myosin or mixed with other
meat proteins (as occurred during the manufacture of the
restructured steakette). When the sarcoplasmic proteins were the
only proteins present in the binding matrix there was virtually
no binding ability at all.

In summary, the sarcoplasmic proteins do affect binding in
meat systems by altering the ionic environment and interacting
with other meat proteins. However, changes observed by
researchers investigating the effect of sarcoplasmic proteins on
binding may have been produced by changes in ionic strength. The
sarcoplasm contains relatively high concentrations of both
organic and inorganic salts which produce an ionic strength of
0.26.[37] The recent work of Trout[38] has shown that 90-95% of the
variation in binding strength in meat products can be explained
in terms of changes in ionic strength and pH.

Myofibrillar proteins

The presence of salt extractable myofibrillar proteins has
been shown to be necessary for satisfactory binding in both
emulsion and sectioned and formed meat products. Work has been
carried out to determine the role of the individual myofibrillar

proteins in binding. Much of the initial work in this area was carried out by Fukazawa et al.,[39],[40] Samejima et al.[41] and Nakayama and Sato[42],[43] using the individual isolated myofibrillar proteins in model gelation systems. It was generally concluded by these workers that myosin and actomyosin were the proteins that produced the greatest gel strengths and therefore were the most important in binding. In addition they found, in most cases, that actomyosin was a more effective binding agent than myosin.

In contrast to these results MacFarlane et al.,[35] Ford et al.[36] and Turner et al.[10] found that myosin was superior to actomyosin in binding meat pieces together in both a model binding system and in a reformed beef product. Although this difference may seem hard to reconcile, a possible explanation may be found in the work of Yasui et al.[28] Using a model gelation system, they showed that the addition of myosin to actomyosin produced a gel that was much stronger than either myosin or actomyosin when used separately. Hence, the results obtained by MacFarlane and his group may be explained by the interaction of the added myosin with the actomyosin present in the surface of the meat to form a strong binding matrix, and the inability of actomyosin alone to achieve a similar effect.

Although the myofibrillar proteins as a group are known to be the major contributors to binding in meat systems, it is difficult to single out the contribution of the individual proteins. This is due to the fact that the different proteins interact with each other when binding[28] and that the method of extraction and purification of the proteins can have a profound influence on their binding ability.[29]

Stromal proteins

The presence of stromal proteins in meat products is essential as products made without them are soft, jelly-like and lack cohesion.[44] In contrast, meat high in connective tissue is rated as being a very poor quality binder. Obviously there is a level in between these two extremes that forms products with acceptable binding characteristics. In addition to the actual concentration of the stromal proteins, the mechanical treatment they receive, the presence and concentration of added salts, and the temperature of processing all have an effect on the binding strength of the final product.

To establish the role of stromal proteins in meat products Puolanne and Ruusunen[45] isolated thick epimysial tissue from young bull skeletal muscle. They then investigated the effect of pH, salt, phosphate, added water and cooking temperature on the amount of water bound by the proteins and the strength of the gelatin gel produced when the released liquid was cooled. The concentration of salt (in the presence or absence of phosphate) and the concentration of phosphate had little, if any, effect on the amount of water bound and the gel strength of the released liquid. The pH and temperature of processing had a strong effect. Over the pH range 5.0-8.0 both the amount of water bound and the gel strength of the liquid released decreased with increasing pH. However, in the pH range of most processed meat products (pH 5.5-6.5) these changes were small and inconsistent. The most dramatic effect of increasing the processing temperature (from 50°C to 90°C) on the amount of water bound occurred at two temperatures, 60°C and 90°C. At 60°C there was a large increase (33%) in the amount of water bound, due to the swelling of the connective tissue.[46] At 90°C the amount of water bound by the stromal proteins had dropped to its lowest level. This was most likely due to inability of the remaining non-solubilised proteins to bind as much water as those that had already been solubilised by heating to the higher temperature. The solubilised proteins were removed before the amount of water bound was measured. The gel strength of the liquid released increased with increasing temperature, due most probably to the increasing amount of protein solubilised.

Early work by Puolanne and Ruusunen[45] found that adding connective tissue (thick epimysial tissue from young bulls) to emulsion products caused only a slight increase in water binding capacity but a sharp increase in firmness. In a similar study Sadowska et al.[47] looked at the effect of increasing the connective tissue level from 5% of the total protein (the normal level for fat free, visual connective tissue free muscle) to 10% and 15%. Increasing the level of added connective tissue to 100% and 200% of normal level produced respectively a 5% and 12% increase in cooking loss and reduced the firmness by 10% and 23% respectively.

The results from these two investigations tend to contradict each other. This may be due to the different processing temperatures used (70°C for the first study and 80°C for the second) with an additional variable being the degree of comminution, which was not stated precisely by either group. Both of these variables have been found by Puolanne and Ruusunen[45] to affect the product texture and yield.

CONCLUSIONS

Meat components perform many functions in processed products. Intact muscle pieces, fibres and myofibrils contribute directly to texture. Texture is also affected by the inclusion of connective tissue and adipose tissue. The inclusion of air, water droplets and melted fat within the product affect the texture of the cooked material. Ionic environment, mechanical treatments, temperature, pH and rigor affect the release of meat components from tissue structure.

Proteins released from the tissue structure act to form a heat setting matrix and coat other structures. The myofibrillar proteins contribute to heat setting matrix formation in appropriate ionic environments. Sarcoplasmic and myofibrillar proteins may act to coat structural components.

The species, composition and anatomical location affect meat ingredient functionality. Frozen storage, fat oxidation, refrigerated storage, enzymic degradation and high temperatures can decrease meat ingredient functionality. Selection of fresh, properly chilled meats for manufacture greatly aid in the production of high quality meat products.

REFERENCES

1. G.R. Schmidt, R.F. Mawson and D.G. Siegel, Food Technol., 1981, (May), 235.
2. P.G. Schnell, D.V. Vadehra and R.C. Baker, Can.Inst.Food Sci.Technol.J., 1970, 3 (2), 44.
3. D.V. Vadehra and R.C. Baker, Food Technol., 1970, 24, 766.
4. J.C. Bard, "Proceedings of the Meat Industry Research Conference", American Meat Institute Foundation, Arlington, VA, USA, 1965, p.96.
5. T.A. Gillett, D.E. Meiburg, C.L. Brown and S. Simon, J.Food Sci., 1977, 42, 1606.
6. L.W. Solomon and G.R. Schmidt, J.Food Sci., 1980, 45, 283.

7. R.L. Saffle and J.W. Galbreath, Food Technol., 1964, 18, 1943.

8. J.C. Acton and R.L. Saffle, Food Technol., 1969, 23 (3), 367.

9. A. Awad, W.D. Powrie and O. Fennema, J.Food Sci., 1968, 33, 227.

10. R.H. Turner, P.N. Jones and J.J. MacFarlane, J.Food Sci., 1979, 44, 1443.

11. R. Hamm and E.J. Grabowska, Die Fleischwirtschaft, 1979, 58, 1345.

12. L. Kotter and A. Fischer, Die Fleischwirtschaft, 1975, 3, 365.

13. D.G. Siegel and G.R. Schmidt, J.Food Sci., 1979, 44, 1129.

14. C.J. Randall and P.W. Voisey, Can.Inst.Food Sci.Technol.J., 1977, 10 (2), 88.

15. J.C. Acton, J.Food Sci., 1972, 37, 244.

16. D.M. Theno, D.G. Siegel and G.R. Schmidt, J.Food Sci., 1978, 43, 483.

17. A.J. Miller, S.A. Ackerman and S.A. Palumbo, J.Food Sci., 1980, 45, 1466.

18. D.G. Siegel, D.M. Theno, G.R. Schmidt and H.W. Norton, J.Food Sci., 1978, 43, 331.

19. K.H. Koo, Dissertation Abstracts International, 1980, B40 (10), 4726.

20. D.M. Theno, D.G. Siegel and G.R. Schmidt, J.Food Sci., 1978, 43, 493.

21. E.R. Maesso, R.C. Baker, M.C. Bourne and D.V. Vadehra, J.Food Sci., 1970, 35, 440.

22. F.H. Pepper and G.R. Schmidt, J.Food Sci., 1975, 40, 227.

23. R.J. Krause, H.W. Ockerman, B. Krol, P.C. Moerman and R.F. Plimpton Jr., J.Food Sci., 1978, 43, 853.

24. R.G. McGowan, "Method of Preparing a Poultry Product", US Patent 3,503,755, 1970.

25. J.H. Mahon, "Proceedings of the Meat Industry Research Conference", American Meat Institute Foundation, Arlington, VA, USA, 1961.

26. J. Grabowska and Z.E. Sikorski, Lebensm.Wiss.u.Technol., 1976, 9, 33.

27. M. Ishioroshi, K. Samejima and T. Yasui, J.Food Sci., 1979, 44, 1281.

28. T. Yasui, M. Ishioroshi and K. Samejima, J.Food Biochem., 1980, 4, 61.

29. D.G. Siegel and G.R. Schmidt, J.Food Sci., 1979, 44, 1686.

30. R. Hamm and F.E. Deatherage, Food Res., 1960, 25, 587.

31. R. Hamm, "The Physiology and Biochemistry of Muscle as a Food", University of Wisconsin Press, Madison, WI, USA, 1966, p.363.

32. J.R. Quinn, D.P. Raymond and V.R. Harwalker, J.Food Sci., 1980, 45, 1146.

33. D.J. Wright, I.B. Leach and P. Wilding, J.Sci.Food Agric., 1977, 28, 557.

34. J.C. Acton and L.H. McCaskill, J.Milk Food Technol., 1972, 35 (10), 571.

35. J.J. MacFarlane, G.R. Schmidt and R.H. Turner, J.Food Sci., 1977, 42, 1603.

36. A.L. Ford, P.N. Jones, J.J. MacFarlane, G.R. Schmidt and R.H. Turner, J.Food Sci., 1978, 43, 815.

37. J.R. Bendall, J.Sci.Food Agric., 1954, 5, 468.

38. G.R. Trout, "The effect of phosphate type, salt concentration and processing conditions on the binding in restructured beef products", M.S. Thesis, Colorado State University, Fort Collins, CO, USA.

39. T. Fukazawa, Y. Hashimoto and T. Yasui, J.Food Sci., 1961, 26, 541.

40. T. Fukazawa, Y. Hashimoto and T. Yasui, J.Food Sci., 1961, 26, 550.

41. K. Samejima, Y. Hashimoto, T. Yasui and T. Fukazawa, J.Food Sci., 1969, 34, 242.

42. T. Nakayama and Y. Sato, Agr.Biol.Chem., 1971, 35 (2), 208.

43. T. Nakayama and Y. Sato, J.Text.Studies., 1971, 2, 75.

44. J. Schut, "Food Emulsions", Marcel Dekker, New York, 1976.

45. E. Puolanne and M. Ruusunen, Meat Sci., 1981, 5, 371.

46. J.C.W. Chien, J.Macromol.Sci.Revs.,Macromol.Chem., 1975, 12, 62.

47. M. Sadowska, Z.E. Sikorski and M. Dobosz, Lebensm.Wiss.u. Technol., 1980, 13 (5), 232.